T0398289

Evaluating IT Projects

Project management disciplines have been a part of IT for many years. Why, then, are so many challenges still directly associated with how a project is managed? Many projects fail for myriad reasons; most failures, however, stem from poor or inadequate project evaluation and performance appraisal, while improved project planning and direction is considered to be one of the key factors to IT project success. Eriona Shtëmbari arranges evaluation methods and techniques into three groups: managerial, financial and development. This book explores the process of project evaluation and the purposes of evaluation, given its strong relationship to the success of the project. It examines IT project evaluation, identifies methods and techniques to be used throughout the project life cycle, examines the benefits of project evaluation, and proposes a systematic approach/framework of project evaluation to serve as a tool for successful project management.

Shtëmbari analyses the most up-to-date research relating to the process and methods/techniques of project evaluation throughout the project life cycle. From the systematic literature review, she identifies the most usable methods and techniques in project evaluation and focuses on the adequacy of these methods and techniques in the service sector. The theoretical underpinning of the book serves as a base to interpret the interviews in the case study and build a theory as to how the project evaluation context relates to the proposed scientific theory. The findings in this book provide solutions for practitioners to help them boost the evaluation framework and consequently improve their IT project management.

Eriona Shtëmbari has worked for the Ministry of Economy, Trade and Energy in Albania, Chemonics International, MCATA Albania, the Professional Academy of Business, and currently works for the University of New York Tirana, Albania.

Routledge Focus on Business and Management

The fields of business and management have grown exponentially as areas of research and education. This growth presents challenges for readers trying to keep up with the latest important insights. Routledge Focus on Business and Management presents small books on big topics and how they intersect with the world of business.

Individually, each title in the series provides coverage of a key academic topic, whilst collectively, the series forms a comprehensive collection across the business disciplines.

ISSN: 2475–6369

For a complete list of titles in this series, please visit www.routledge.com/business/series/FBM

The Spartan W@rker
Konstantinos Perrotis and Cary L. Cooper

Writing a Business Plan: A Practical Guide
Ignatius Ekanem

Manager vs. Leader: Untying the Gordian Knot
Robert M. Murphy and Kathleen M. Murphy

Accounting for Biological Assets
Rute Goncalves and Patricia Teixeira Lopes

Rising Consumer Materialism: A Threat to Sustainable Happiness
Afia Khalid and Faisal Qadeer

Evaluating IT Projects
Eriona Shtëmbari

Evaluating IT Projects

Eriona Shtëmbari

Routledge
Taylor & Francis Group

LONDON AND NEW YORK

First published 2018
by Routledge
2 Park Square, Milton Park, Abingdon, Oxon OX14 4RN

and by Routledge
711 Third Avenue, New York, NY 10017

Routledge is an imprint of the Taylor & Francis Group, an informa business

© 2018 Eriona Shtëmbari

The right of Eriona Shtëmbari to be identified as author of this work has been asserted by her in accordance with sections 77 and 78 of the Copyright, Designs and Patents Act 1988.

Trademark notice: Product or corporate names may be trademarks or registered trademarks, and are used only for identification and explanation without intent to infringe.

British Library Cataloguing-in-Publication Data
A catalogue record for this book is available from the British Library

Library of Congress Cataloging-in-Publication Data
A catalog record for this book has been requested

ISBN: 978-1-138-10411-2 (hbk)
ISBN: 978-1-315-10232-0 (ebk)

Typeset in Times New Roman
by Apex CoVantage, LLC

With love to my precious children Alisa and Lukas

Contents

Illustrations

Figures

Tables

Preface

A critical issue in service projects, especially IT projects, is how to evaluate their success from the start. This book explores the process and identifies the methods for evaluating projects from the initiation phase to close-out, by illustrating them with a case study to show how they work in the business environment. Many projects fail for myriad reasons; most failures, however, stem from poor or inadequate project evaluation and performance appraisal. The literature reveals that many studies have been discussing the issue of IT project selection and associated investment decisions, but very few authors have considered how organizations settle on project evaluation, which fosters the main source of success. The aim of this book is to increase the chances of success through the use of a systematic evaluation process and appropriate methods and techniques. The goal is to improve overall project management, especially in the IT sector.

Moreover, evaluation should be performed throughout the IT project life cycle and not only at the first stage of project selection. This book has grouped evaluation methods and techniques into three groups: managerial, financial and development, under a strategic approach of decision-making upon the service projects. Empirical data are gathered through a case study conducted in Bank X in Albania.

Finally, by combining both findings from systematic review of the existing literature on this topic and findings from the empirical study, the book provides a framework for guiding practitioners on how to evaluate their projects and successfully launch innovative services.

This book will:

1 help practitioners understand the process of project evaluation and the purposes of evaluation given its strong relationship to the success of the projects;
2 help practitioners examine IT project evaluation and identify methods and techniques to be used throughout the project life cycle;

3   propose a systematic approach/framework of project evaluation, which
    can serve as a tool for successful project management; and
4   add to existing body of knowledge from both points of view – academ-
    ics and practitioners.

Acknowledgements

It is a pleasure to thank all the kind people around me who helped and supported me through the challenging journey of writing this book.

Above all, I owe my deepest gratitude to Professor Corrado Cerruti for his precious advice throughout this research process, which kept me focused on the topic and improved the quality. Without his continuous optimism, enthusiasm, encouragement and support concerning this work, this study would not have been possible. His professionalism is highly appreciated.

I am truly indebted and thankful to the respondents of Bank X for their availability and information sharing. Without their support, the empirical study of this research would not have been possible.

A special thank you to my husband Gert, and my precious children, Alisa and Lukas Joel, for their personal support, love and great patience at all times. My parents, brother and extended family have given me their unequivocal support throughout, as always, for which my mere expression of thanks likewise does not suffice, and without them I wouldn't be where I am today. Finally, I express my deepest appreciation to my mom, who prays for me every day.

Introduction

This book explores the up-to-date research done in project evaluation throughout the project life cycle and examines practical insights from the banking sector, particularly IT projects. The author looked at the evaluation purposes, process and evaluation methods/techniques. All are integrated in a framework for successful project evaluation, which consist in the main contribution of this book.

Project evaluation is an essential component of the project management process that often tends to be forgone or neglected. Performed regularly, evaluation fosters early problem detection and correction by enhancing feedback and generating continuous improvement in the project management process. The evaluation provides feedback information to the project manager, concerning success or failure of the project activities, for their decision-making function. As discussed by many authors, evaluation becomes not only an instrument but also an irreplaceable source of information for continuing planning where goals and objectives are improved. In an evaluation process it is desirable to use methods strong enough to detect any project efforts but at the same time would not respond significantly to any change not caused by project activities.

The project evaluation process is, therefore, carefully undertaken during the project life cycle by organisations in order to ensure that the project is profitable, that it is on the right track with expected parameters and that the goals of organisations would be achieved once the project is completed. Many companies report that they are uncertain how to measure the performance of their projects. In 2001, the performance evaluation of new product/ service developments was surveyed in some companies in the UK by Driva, Pawar & Meno. Without exception, all companies wished to improve their use of performance evaluation. This implies that the methods and techniques used by these companies were not satisfactory. Estimates of the project's effects depend on the evaluation method (Clemens and Demombynes, 2011). Evaluation frameworks are increasingly required from practitioners

(Ahlemann et al., 2013). Surprisingly, academic research in this area, until recently, has been rather limited.

The economic landscape is being restructured by recent financial crises, by shifting the attention to the service sector as the basic mode of generating revenue today (Huff and Möslein, 2009). Moreover, given the intangible traits of services, the evaluation of service projects becomes more difficult and complicated, leaving room for further research, and thus leading the author to study this sector. Furthermore, 'according to IT project management statistics, every two out of three IT projects fail partially or totally. Yet, although many organizations have adopted good project evaluation processes and some project managers are undoubtedly learning from evaluations experience, the failure rate does not seem to be decreasing' (Nelson, 2007). This lack of statistical improvement may be due to the rising size and complexity of projects, the increasing dispersion of development teams and the reluctance of many organizations to perform project evaluation. Besides, project and program evaluators have paid little attention in the literature to the quality of implementation, which is very important to decision-makers (Brandon et al., 2014).

This book investigates the IT sector due to the fact that IT projects are, on one hand, becoming more and more important to the organisations, while on the other hand they have the highest risk of failure, under intense scrutiny by C-level executives looking to cut costs. As a result, project management becomes more important in today's economic conditions. IT project managers need to consider and understand how the downturn may affect their operations and then take appropriate actions. Poor estimates or inappropriate evaluation methods used to define requirements at the project planning stage also contribute to project failure. A report from Wiklund and Pucciarelli (2009) states that 25 percent of IT projects fail outright, while 20 to 25 percent don't provide ROI and up to 50 percent require material rework. Also, 'a decision to continue with a bad plan may lead to a failed project, whereas requesting unnecessary additional planning for an already high-quality plan may be counterproductive' (Féris et al., 2017).

Project management disciplines have been a part of IT for many years. So why are so many challenges still directly associated with how a project is managed? The issues often boil down to several reasons for project failure. Improving project planning and direction is one of the key factors in IT project success. This requires adequate information to the decision-makers.

This book highlights that appropriate evaluation throughout the IT project life cycle should improve the evaluation decisions and lead to more effective and efficient use of IT resources. Effective planning and management are crucial to the conduct of a successful evaluation (Gudda, 2011).

The expectation is that there will be an improved targeting and more strategic use of IT resources resulting in a positive impact, either directly or indirectly, on the overall profitability of the company.

The outcomes of the research done in this book were findings on the methods/techniques and the process which enables several benefits among which the most important is adequate information for decision-makers, with a focus on service and IT projects. Moreover, the author wanted to compare the methods and techniques proposed by literature with those used by Bank X for project evaluation. And finally, the author wanted to understand the framework in which these methods and techniques are used during the whole project life cycle. The theoretical base served as a base to interpret the data collected through interviews, guidelines, worksheets and templates of Bank X, to build a theory on how the project evaluation context relates to the proposed scientific theory. Findings of this study aim to help practitioners boost the evaluation framework by choosing adequate methods/techniques and consequently improve their IT project management.

The study presented in this book is being conducted with a hope to facilitate the evaluation process in launching innovative IT projects.

Organization of the book

In order to provide readers a clear and logical approach to the topic, the book is divided into three chapters as following:

Introduction sets the **general** research interest and background of the study and explains reasons for choosing the research subject.

Chapter 1 – **Project Evaluation** provides specific and concrete review of the project evaluation field. It gives insights to the concept, process and different purposes of project evaluation and summarizes important factors that influence project evaluation.

Chapter 2 – **IT Project Evaluation** is focused on IT project management analyzing its characteristics and main factors that influence their evaluation. Empirical evidence is brought from the banking sector regarding the process and purposes of evaluation. Finally, challenges met in IT project evaluation are also brought to the reader's attention.

Chapter 3 – **Methods and Techniques of Evaluation** brings valuable discussion on the general evaluation methods and techniques throughout the project life cycle. The focus then moves to the analyses of methods and techniques used to evaluate IT projects. Finally, the evaluation framework is provided for a successful evaluation of IT projects.

1   Project evaluation

The *APMBoK and the PMBOK® Guide* are considered as standard author-
ized guidelines that support the project management domain. The *APMBoK*
section on evaluation and control, which includes many of the traditional
tools associated with project evaluation, emphasizes the importance of project evaluation and control during the project life cycle. This is an important
difference between the *APMBoK* and the *PMBOK® Guide.*

Project and Program evaluation are used as synonyms in this book. Both
terms refer to the evaluation process of projects having starting and ending
points. Moreover, different terms are used to express same results from a
process of evaluation. Terms such as evaluation, appraisal, review, monitoring and control are used as synonyms in the literature review.

1.1 The process of evaluation throughout
the project life cycle

The process of evaluation is an important tool in project management; therefore, it has been a subject of study from many authors. One of the key concerns of managers is the control and evaluation of the overall development
effort (Nurmi et al., 2011) and (Mertens and Wilson, 2012), although there
are problems before the process has even begun (McAvoy, 2006). However,
project reviews should add value not only to future projects, but also to the
current project (Highsmith, 2004).

Gramham (2006) argues that it is impossible to set meaningful targets
for profitable project outcomes, without appropriate measurement and
evaluation systems in place. Reliable evaluation techniques and criteria
are becoming more and more important to stakeholders who are interested
either in a specific project or overall activity of the company (Akalu, 2003;
Oral et al., 1991).

The successful performance of a project depends on appropriate planning. The *PMBOK® Guide* defines the use of 21 processes that relate to

planning out of the 39 processes required for proper project management (Globerson and Zwikael, 2002). Execution of the project according to the predefined project plan can be achieved through a project evaluation methodology. Consequently, project evaluation is a significant issue during the project life cycle. The design of a project evaluation system is an important part of the project management effort (Shtub et al., 2005).

In addition, project success is not only determined on the basis of the three traditional perspectives – time, cost and quality – but also on long-term benefits, continuous improvement and sustainability of the project's outcomes. Despite the continuous evolution in the project management field, it appears evident that traditional approaches still show a lack of appropriate methodologies for project evaluation and control (De Falco and Macchiaroli, 1998). Many articles have supported the importance of project evaluation in the achievement of the project aims and objectives as well as decision-making process. Project performance can be improved if more attention is given to the issue of evaluation (Avison et al., 2001). Evaluation means judging; appraising; determining the worth, value or quality of the project to make necessary decisions in terms of strategic project management impact, effectiveness and efficiency.

Project evaluation is a combination of a number of activities (Steven et al., 1993). Moreover, the evaluation process concerns very much on data gathering and information analysis. Therefore, the sources of information, the reliability and validity of feedbacks that contribute to decision-making in program evaluation, become crucial (McNamara, 1994; Dura et al., 2014). Liang (2003) and Lagsten (2010) emphasize the critical role of management support, emphasized also by, taking into consideration the novelty of the project. This book argues that understanding the top management perspective on evaluation decision-making is a critical issue of project evaluation, because it is widely recognized that senior management commitment is a key ingredient to producing successful IT project outcomes. Further, comparing and contrasting the top management and PM perspectives on evaluation process may yield insights into how PMs can forge better relationships with top management. This would seem to be important, as the literature suggests that one of the ways in which PMs can improve project performance is to develop strong relationships with top management (Simonsen, 2007).

Another practical evaluation process focuses on how to establish indicators looking at both internal and external context of projects (Örtengren, 2004; Kaufman, 2014).

During the late 1960s, Michael Scriven coined the terms summative and formative evaluation. Since then, new roots of evaluation have been growing, engaging a growing number of people which use a wide variety of tools and techniques to support project evaluation and monitoring. Steven et al. (1993) categories project evaluation into three phases which are planning phase, formative phase and summative phase. Later, McNamara (1997) expresses this consideration in another way with three major types of project evaluation: goal-based, process-based and outcome-based evaluation. While practitioners refer usually to the project evaluation phases as: *investment appraisal, performance evaluation, and end project review. We will refer to these terms during our study in this book.*

Let us discuss now each of the evaluation phases in details, illustrating them with empirical evidence from the case study.

1.1.1 Investment appraisal phase: goal-based

This type of evaluation occurs prior to the project and is useful in selecting the appropriate project. It serves to evaluate benefits, risk and costs associated with the project. Several methods have been proposed to help organizations make good project selection decisions (Stewart and Mohamed, 2002; Klapka and Pinos (2002); Osman et al., 2005). However, many reported methods have several limitations and tend not to provide a means to combine tangible and intangible business value and risk criteria. Others are too complex in structure and have little appeal to practitioners. The methods and techniques of project evaluation will be discussed in detail in Chapter 3 by giving hints as to which one is more appropriate to use in each phase of the project life cycle and also illustrating in practice through the chosen case study.

1.1.2 Performance evaluation phase: process-based

The second phase deals with the implementation and the ongoing evaluation and monitoring of the project. Within most sectors of government and private industry there are suggestions that project investments are often accompanied by poor vision and implementation approaches, insufficient planning and coordination and are rarely evaluated throughout the project life cycle. The successful implementation of new and innovative projects requires the development of evaluation plans prior to project commencement (Pena-Mora et al., 1999). Effective evaluation during the planning phase should go some way to reduce the current gap between output and expectation from project investments (Dos Santos and Sussman, 2000). Recently, the interest in initial evaluation for proper planning frameworks to aid project implementation has grown.

1.1.3 End project review phase: outcome-based

This is the last phase dealing with project closure. Generally, service project investment appraisal is more difficult than other investment decisions, especially in the IT sector, because IT-induced benefits are hard to identify and quantify (Stewart and Mohamed, 2003).

The process of project evaluation runs through several steps including different activities in each stage. These activities include project objectives and potential influential context assessment, indicators establishment, progress analysis, and 'actual versus estimated parameters' comparison. Among those activities, setting indicators is considered one of the crucial activities of evaluation. The argument behind this reasoning is that the project evaluator has to refer to the initial criteria to keep the project on the right track and to move the project toward the common goal all the time; judicious evaluation criteria set at the beginning enhances the success possibility of the evaluation process. Beside, most respondents mentioned collecting information through communication (Oral et al., 1991; Steven et al., 1993; McNamara, 1994; Misso Kwon et al., 2010; Morris and Pinto, 2010) as a big part of evaluation process. The feedback and feed-forward channel is therefore of special concern to both academics and practitioners during the evaluation process. In general, the model collected from practices shows more details on each activities of evaluation process, which turn out to give a clearer idea of what should be done, who should do and when should the evaluation be done. Undertaking a proper project evaluation throughout the project life cycle would enable organizations to achieve desired objectives within time, cost and quality requirements.

To conclude, even though there are some differences in the way of undertaking evaluation, literature on project evaluation, from both authors and practitioners, emphasizes the importance of stakeholder involvement, of information quality and of criteria establishment in the entire process. Moreover, the project evaluation is mentioned by both authors and practitioners as an iterative process which goes along all phases of the project life cycle.

Project evaluation landscape in Bank X

The case study revealed that there are two main groups of people evaluating IT projects in Bank X. These two groups are:

Group 1: **Project Management Office** (PMO) and more precisely the **Project Officer** (PO). All projects are to be monitored by the PMO in the Organization and Project Management Department (OPMD) as a central unit, which keeps track of the bank projects portfolio. The PO supports the Project Manager in each project phase and assures Project Management Methodology and Standards implementation. Moreover, he monitors the projects with regard to time, cost and quality of projects, including a verification of reported figures.

Group 2: **Project Manager** and **Project Team**

The **Project Manager** should prepare precise, appropriate, realistic and attainable project planning. The Project Manager is the person responsible for defining scope, time and budget in the very beginning and for evaluating the accomplishment of the project objectives within scope, time and budget. He or she should reflectto the relevant documents every change of scope, time and budget. The Project Manager has the responsibility of cost tracking during the project life cycle. On project completion the expected benefits of the project should be tracked and evaluated according to project methodology.

The **Project Team** consists of anyone who is required to contribute to the project by carrying out the work, e.g., representatives of business, operations, risk, IT area, information security, etc. While the Business Project Team members are responsible for all input from the business side on the basis of their special expertise, especially on defining the business requirements, the IT Project Team members have to transform this input into IT concepts and finally IT results. They are also responsible for all IT documentation tasks. The information security member shall perform the security risk analysis and in cooperation with business and IT decide on the set of controls necessary to be implemented in the solution.

Both groups report to the **Steering Committee** (StC), which is typically composed of stakeholders whose cooperation and support is a prerequisite for project success. These stakeholders are likely to be in a position to influence the project progress and have the resources required by the project in terms of people, budget and equipment. Further members of the StC are key stakeholders – department heads affected by the project and responsible board members or division heads if the project is a strategic one. The StC is responsible for all decisions about scope, change requests and authorizations of the project during its delivery phase and closure, and if the decision does not influence other projects.

Project prioritization is carried out by the first evaluation group: the Project Officer in the PMO, prior to the project implementation to justify the decision to develop the new service. The PMO regularly prepares a list of new projects proposed by placing a high priority on running projects that have continued. Several rounds of prioritization with line board members

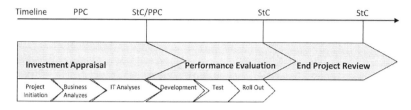

Figure 1.1 Evaluation process in IT projects

and IT are organized in order to define a realistic project portfolio. These results are reflected in a better project evaluation and cost tracking on yearly basis in Bank X's project portfolio.

Once the project is approved, the project manager is assigned and leads the project team to develop new service. During that period, the project manager and the project team continues to evaluate the progress and performance of the project and report to the StC on a weekly or monthly basis. After the official launch of the new service, the Project Manager and the Project Officer conduct another evaluation to see what went wrong, what went right and document all the solutions for the next project improvement as well as for organizational knowledge transfer. Figure 1.1 illustrates the overall process of IT project evaluation throughout the project life cycle.

DAILY EVALUATION PROCESS FOR FIXING DAILY BUSINESS DEFECTS OR
SMALL CHANGES IN PROJECTS

In case business requires repairing a small defect/problem or a change that occurs during the daily business in a live environment, he/she should send a signed request 'Statement of Work' to IT QA and the Account Management Department. IT will evaluate the request and the required time to fix the defect/problem.

In case the estimated time will not impact the project portfolio and requires less than ten man days (MD)MDs for local IT, IT will take care to fix every defect/problem and deliver the original signed Statement of Work to the PMO.

In case it comes with high priority and requires less than ten MDs, but it impacts the project portfolio, the PMO should be informed for such impact in order to ask the Project Portfolio Committee (PPC) for prioritization.

In case the estimated time for a request is ten to 20 MDs, Board Member approval is needed.

In case the required time is estimated to be more than 20 MDs and impacts or not the project portfolio, the requestor will be notified to go through PMO and screen the request as a new initiative.

When changes are required in the frame of an existing project, the Project Manager, in collaboration with the Project Team, will prepare 'Change Request' form and deliver it to IT after it is approved as a change of scope by the Sponsor. If the estimations (time and costs) are accepted, the project Sponsor and StC will approve the updated project plan. In case the required time impacts the project portfolio and/or project milestone timelines, the requestor will be notified to go through PPC and ask for prioritization. In case the required time does not impact the project portfolio and/or project milestone timelines, IT will develop the required change request and send the documentation to PMO to file it with the project documentation.

1.2 The purposes of project evaluation

It is found from literature that the evaluation process plays an important role in the success of projects. One of the major purposes of evaluation is to determine the worth or merit of projects, process or products under both internal and external constraints. According to Ye and Tiong (2000), financial appraisal techniques in evaluating projects offer quantitative information to justify investment, particularly in a high level of finance and political risk projects such as infrastructure. Chapman et al. (2006), on the other hand, argued that evaluation as a risk management tool allows project managers to reduce uncertainties when making decisions, especially before the project started.

Moreover, they recognize that there is an iterative relationship between the evaluation process and decision-making and the subjective opinions of evaluators on the assumptions of the project that might lead to irrational decision. Hence, by combining both monetary and non-monetary aspects, project managers aim to choose the most appropriate projects and make determined decisions on the projects while doing evaluation.

Frechtling (2002) mentions two reasons for conducting evaluation: (1) 'it provides information to help improve the project, information on whether the goal are being met and on how different aspects of a project are working are essential to the continuous improvement process; and (2) 'provides new insights or new information that was not anticipated' (p. 11). He also stresses the role of evaluation process in facilitating information flows among stakeholders of organization. It enhances the feedback and feed-forward mechanism through reports and questions that are delivered along the evaluation period. Caulley (1993) agrees on this point and suggests that evaluation could provide objective information, supply credible answers, and identify the reasons for success or failure.

The insight of the project progress achieved through the evaluation process allows management to take proper actions during the implementation of projects. Banwell et al. (2003) further illustrates the role of project evaluation as a toolkit that 'helps managers to guide and benchmark the development of organizations' in relation to the adoption of the project. In general, evaluation gives the initial basis for monitor and control procedure of organizations.

Gudda (2011) argues that valuation maybe categorized in two kinds of stages: formative evaluation and summative evaluation. The purpose of formative evaluation is to assess whether the project is being conducted as planned, while the purpose of summative evaluation is to assess a mature project's success in reaching its stated goals.

However, Brown and Remenyi (2002) comment that there is a growing consensus among academics that various ranges of issues should be addressed during evaluation. The traditional view of financially driven measurement-oriented evaluation should be replaced by a form of evaluation which is concerned with ongoing learning rather than simply measuring. In addition, the International Development Research Centre (IDRC) argues that evaluations shift away from a tool for 'control' to a tool that can empower organizations and contribute to organizational learning. Evaluation is also considered as knowledge construction and capacity-building by some authors (Vakola, 2000; Segone, 1998).

The APM body of knowledge (2006) discusses that project evaluation review should take place to check the likely or actual achievement of a project's plan and to ensure the benefits of organization. Agreed with previous academic literature, it is also emphases that evaluation should be undertaken throughout the project life cycle.

To support this opinion, Farbey et al. (1992) summarizes that evaluation: (1) could be used as part of justification for a project, either an existing or new; (2) enables organizations to compare numbers of projects under constraint of resource available resulting in judicious decisions; (3) provides a set of measures supporting the monitor and control system; and (4) determines the success or failure of projects based on initial benchmarks and provides lessons learned for the future (p. 110). An empirical study on 16 investments on IT project in various industries prop up the findings.

Despite touching different aspects of project evaluation, all the literature concurred on the embeddedness of project evaluation on the decision-making process along the project life cycle. It aims at problem solving and decision-making (Sherwood-Smith, 1994, in Vakola, 2000; Wholey et al.,

1994; Scriven, 1980). Evaluation gives criteria for the selection of a project, assesses current project status for the decision whether to continue with the project and, finally, provides suggestion on taking further investment or not.

In order to give a practical insight, the following section will illustrate the purpose of project evaluation using the same case study conducted in a bank.

1.2.1 Empirical evidence from the case study

The case study results showed that the project evaluation is a very important process not only to the decision-makers but to the whole bank.

Project evaluation methods/techniques allow Project Managers to minimize project risk by using structures and principles. This methodology provides instructions for project management and defines general rules and minimum standards to be adhered to.

A well-established project evaluation framework contributes significantly to the overall success of projects and will mainly support the Board of Management in fields of Project Management including Project Portfolio Management. This is mandatory for all Bank X units and its implementation will be regularly audited by the Bank X Internal Audit Department.

Regarding the purpose of evaluation, the evaluation environment in Bank X fosters many reasons for conducting an evaluation, including:

1   It provides managers with information regarding project performance. Project plans might change during the implementation process. Evaluations can verify if the program is running as originally planned. In addition, they provide signs of project strengths and weaknesses, and, therefore, enable managers to improve future planning, delivery of services and decision-making.

2   It assists project managers, staff and other stakeholders to determine in a systematic and objective manner the relevance, effectiveness and efficiency of activities (expected and unexpected) in light of specified objectives.

3   Mid-term evaluations may serve as a means of validating the results of initial assessments obtained from project monitoring activities.

4   If conducted after the termination of a project, an evaluation determines the extent to which the interventions are successful in terms of their impact and sustainability of results.

5   It assists managers in carrying out a thorough review and re-thinking their projects in terms of their goals and objectives, and the means to achieve them.

6   It generates detailed information about the project implementation process and results. Such information can be used for public relations,

fundraising, promotion of services in the community, as well as identifying possibilities for project replication.

7   It improves the learning process. Evaluations often document and explain the causes as to why activities succeeded or failed. Such documentation can help in making future activities more relevant and effective.

1.3 Key factors influencing project evaluation

In order to successfully implement the project, it is important to make sure the project is in alignment with the overall strategy of the company. Therefore, criteria measuring the compatibility and consistency of a project with company's strategy and long-range plan (Twiss, 1986; Chiesa and Masella, 1996) are seen as key factors of project evaluation.

Project and programme personnel are often 'suspicious of the motives of evaluators' (Grabe, 1983). Therefore, it is crucial to establish conditions of trust between project staff and evaluation personnel. Moreover, 'it is necessary to draw clear lines of distinction between monitoring and evaluation' on the one hand, and 'other management control functions such as auditing, periodical reviews and others' on the other hand (Grabe, 1983). Grabe further argues that there are some criteria to be taken into consideration while evaluating the project: 'First, project design and implementation must be relevant and remain relevant throughout the period of implementation. Second, action taken under the project should be effective. Third, it should have a significant impact, producing lasting change. Finally, it should be efficient, which means that it should produce the desired results, with a minimum of undesirable side-effects, at lowest possible cost (a high cost/effectiveness ratio)' (Grabe, 1983).

The author has grouped main evaluation factors found in different studies, as shown in Table 1.1.

Table 1.1 Project evaluation factors

Sources	Criteria of PE
Örtengren (2004)	Factors of success mostly mentioned in evaluation of project/program: • Commitment of all parties involved • Division of work/responsibilities • Clear and realistic objectives/goals • Specific links between project activities and objectives • Capacity of project group • Flexibility to adapt • Level of participation of end users in project evaluation

(Continued)

Table 1.1 Continued

Sources	Criteria of PE
Andersen et al. (2002)	Project evaluation scheme: • Results of project is potentially of great value to customers • Clear project scope • Alignment with organization's strategies • Involvement of all stakeholders from the beginning till the end of project life cycle • Quality parameter is clear • Financial and technical control • Internal and external communication
Tukel and Walter (2001)	• Time – critical for NPD (introducing idea as soon as possible) • Cost • Quality – critical for NSD • Customer focused (more in NSD) • Rework reduction • Technical specific focused
Millis and Mercken (2004)	• Balance score card in ICT projects • Financial perspectives • Customer perspectives • Internal perspectives • Innovation and learning perspectives
Farbey et al. (1992)	• *Decision environment* (match the culture of the organization) (Evaluation may have to conform to an existing corporate procedure or there may be no established practice.) • *Organization characteristics*: industry situation (stable or not) and leadership role (pioneer or follow) • *Cause-effect relationship*: the degree to predict the impact of new service determines how to do evaluation
Oral et al. (1991)	• Economic contribution • Technological contribution • Social contribution • Probability of success • Resource requirement
Grabe (1983)	• Relationship between project staff and evaluation personnel • Draw clear lines of functions and responsibilities (in monitoring and evaluation – other management control functions such as auditing, periodical reviews and others)

1.4 Implications

The study suggests that top management needs to understand basic criteria related to project evaluation. Top management needs to select the process, choose and customize the methods and techniques to the specific project context/approach and phase of project life cycle.

Lack of support from senior executives continues to represent a major risk – maybe because senior executives themselves do not realize the critical role they can play in helping to deliver successful projects. The author concludes that understanding the top management perspective on evaluation decision-making is a critical issue of project evaluation, because it is widely recognized that senior management commitment is a key ingredient to producing successful IT project outcomes.

2   Evaluating IT projects

2.1 IT project management

Information technology (IT) has become a key business function for almost every organization and most have great expectations of their investment in IT for the future benefits to the business. Yet, the traditional role reserved for IT has been fairly passive in business planning. It has been merely an implementation tool, not intrinsically involved in shaping strategy. Now, as more and more business opportunities and channels to market are created by technology developments, IT plays an increasingly proactive role in developing long-term business objectives. Obviously, managers want to know what went wrong in the hopes they can avoid similar outcomes going forward.

This book stresses the importance of performing an accurate evaluation throughout the life of the project.

The introduction of IBM's personal computer (PC) in 1981 is perhaps the most significant technological achievement of the twentieth century. The PC was further enhanced by the public introduction of the Internet in 1992. Moreover, the portable document file (PDF) created by Adobe in 1993 made it possible to move documents to an electronic format. What this means for evaluators is that there is now a tool that can provide for improved management of electronic files. Digital diaries can be used to record and track evaluations as lessons, which can be stored real time in a repository.

IT has now become an important tool for achieving competitive advantage. IT has indeed become an indispensable element in many firms' business strategies. Business competition is global, intense and dynamic. Information systems and technology (IST) is a key resource in responding to and proacting with this environment. Taking into account the discussion of Stewart (2008), it is obvious that IT is becoming very important and is recently implemented in many organizations for strategic use. Senior executives are increasingly feeling the need to become informed, energized and engaged in information systems. Further, the importance of top management support in IT projects is well acknowledged. Prior research has shown that lack of support from top management is the top risk in IT projects. Surprisingly, senior executives' perceptions of the importance of IT project evaluation have never been systematically examined. One reason why lack of support from senior executives continues to represent a major risk may be that the executives themselves do not realize the critical role they can play in helping to deliver successful projects.

It should be recognized that IT is a strategic source that links information systems to business strategy. IT projects enable firms to adhere to business objectives and to maximize the value from investments.

Despite the huge effort on establishing a suitable framework for IT projects evaluation, most of the work gives few hints on the evaluation of service development projects (Johne and Storey, 1998). Therefore, we felt that a higher contribution is needed on the service industry. This is especially critical due to the increasing contribution of services to the global economy. The service sector has, for a long time, been emphasized for its importance, pointing out that it is the service elements that make the difference on the marketplace and not because of the product components in the manufacturing's offering (such as Grönroos, 1998). Moreover, the service sector counts for over 50% of gross national product or total employment in developed countries (Grönroos, 2000). But, the success of New Service Development (NSD) projects is challenged by specific traits of service product (intangibility, heterogeneity and non-storability), as well as by the novel ideas and the high risk of failure. This makes the evaluation process of NSD projects much more complex and requires special attention. It happens that many projects fail to appeal to intended customers or fail to add value to the organisations' business. Others have been considered as not efficient enough because they are not well evaluated before, during and after the project implementation (Nelson, 2006, Örtengren, 2004).

IT applications nowadays are contributing to improve operating efficiency, create new business opportunities, and even fundamentally altered the rules of the game in competition in certain businesses, thus *becoming a representative case in the NSD sector.*

Business benefits from IT activities come in many forms; not all of them can be measured in financial terms, at least not directly. Of course, business initiatives will be brought forward to achieve real, tangible results, whereas many pure IT projects are unlikely to provide a direct return on investment. Then, *how could we evaluate this kind of project?, Which would be the appropriate technique that would foster adequate information to the decision-makers?*

This issue triggered the author's interest to focus especially in the IT sector. Service projects require a special attention during the appraisal process given their intangible traits. Adoption of IT initiatives can be more lengthy, expensive and complex than other service projects (Gunasekarana et al., 2001).

Many organizations have been increasing their investments in IST to meet the growing demands for efficiency and effectiveness (Gunasekarana et al., 2001). Well planned IST investments that are carefully selected with respect to business mission requirements can have a positive impact on organizational performance (Rivard et al., 2006). Conversely, IST investments that are poorly planned can postpone or severely limit organizational performance and make them counterproductive (Gunasekarana et al., 2001; Féris et al., 2017). 'There is concern that poor evaluation procedures mean it is difficult to select projects for investment, to control development and to measure business return after implementation' (Farbey et al., 1992, p. 189). Often, organizations need to choose between a number of competing IT investments for various reasons including limited resources and capacity constraints.

IT investments also tend to have a high failure rate that might have potentially devastating impacts (Wu and Ong, 2008). One of the most critical characteristics of IT investments that differentiates them from the other types of innovative projects is the high degree of risk and uncertainty associated with them (Bacon, 1992; Gunasekarana et al., 2001; Irani et al., 2002; Wu and Ong, 2008). The complexity of IT projects as well as their interdependencies poses a challenge when applying methods for the prioritization of these investments (Bardhan et al., 2004). That is why authors have criticized conventional methods for evaluating IT investments (e.g., Farbey et al., 1993). Therefore, it is necessary to deploy to a process through which the projects should be evaluated in a comprehensive way considering all special factors having an impact on IT projects (Misso Kwon et al., 2010).

2.2 Key factors influencing IT projects

There are numerous and classic mistakes influencing IT projects. One approach is to group them into the four categories of people, process, product and technology (McConnell, 1996). A later study conducted by Nelson (2007) concluded that 54% of the projects fail because of poor evaluation and/or scheduling, ranked as the first biggest classic mistake, and 51% fail due to ineffective stakeholders management. We will focus our discussion on these two main factors that comprise the biggest percentage of failure in IT projects: evaluation/scheduling and stakeholder management.

Project evaluation and scheduling: The evaluation and scheduling process consists of sizing or scoping the project, estimating the effort and time required, and then developing a calendar schedule, taking into consideration such factors as resource availability, technology acquisition, and business cycles. An accurate evaluation of the project would result in fewer mistakes; less overtime, schedule pressure and staff turnover; better coordination with non-development tasks; better budgeting; and, of course, more credibility for the project team. Four valuable approaches, according to Nelson (2007), to improving project evaluation and therefore project management include: (1) timebox development because shorter, smaller projects are easier to estimate; (2) creating a work breakdown structure to help size and scope projects; (3) retrospectives to capture actual size, effort and time data for use in making future project estimates; and (4) a project management office to maintain a repository of project data over time. Advocates of agile development contend that their methods facilitate better evaluation and scheduling by focusing on scope, short release cycles, and user stories that help estimate difficulty (rather than time).

Stakeholder management: The importance of stakeholder management in NSD projects is raised by many authors such as by Thomas (1978), Alam and Perry (2002), Shneider and Bowen (1984) and Jong and Vermeulen (2003). Key stakeholders such as senior executives are rarely involved in evaluating IT projects. Based on prior research (Keil et al., 2002), it is important to note that various stakeholders may hold different perceptions regarding project management. Therefore, the key for successful project evaluation is to decide on evaluation methods and techniques, determine the value that evaluation brings and what the stakeholders desire (Thomas 1978). In some instances, this can give rise to serious problems. As Nolan and McFarlan (2005) point out, 'a lack of board oversight for IT activities is dangerous; it puts the firm at risk in the same way that failing to audit its books would'.

The evaluation is a knowledge creating process and should therefore be open for including stakeholders during the ongoing process (Lagsten, 2010). There are some best practices for improving stakeholder management, including the use of a stakeholder worksheet and assessment. This tool facilitates the three-dimensional mapping of stakeholder power, level of interest, and degree of support/resistance (Nelson, 2007). Other best practices in this area include the use of communication plans, creation of a project management office, and portfolio management. Getting top management support for a project has long been preached as a critical success factor, so it was somewhat surprising to see insufficient project sponsorship as a major issue in over one-third of the projects. On the other hand, this finding solidifies its status as a *classic* mistake. In some cases, the support was lacking from the start. In other cases, the key project sponsor departed midstream, leaving a void that was never properly filled.

By taking into account key factors influencing IT projects and delving deeper the project evaluation process, we think that it would be possible to improve the chance of success in IT projects.

2.3   IT project evaluation process throughout the life cycle

Although project evaluation is considered an important process fostering a successful IT project management, there is very little or no uniformity in evaluating IT projects in order to measure risks, benefits, and costs of various IT projects (De Reyck, et al, 2005; Irani, 2005 and Gunasekaran, 2006). For instance, Stewart (2008) has developed module frameworks for project IT and companion software package. However, each phase should not be viewed as a separate step. Rather, each is conducted as part of a continual, interdependent management effort. Information gained from one phase is used to support activities in each of the other two phases.

2.3.1   Empirical evidence of IT evaluation process

The case study revealed that there are two main groups of people doing the project evaluation. Each group is involved in different phases of project evaluation process, generating different benefits. Group 1 (PMO) evaluates the project as a business opportunity while Group 2 (Project Manager and Project Team) evaluates the performance and progress of the project. Therefore, the PMO and Steering Committee start evaluating projects from the first phase until the last phase whereas, the project manager only starts assessing the project from the second phase and together with their StC look back to gain lessons in the last phase of project evaluation. While the PMO's role in the first phase is crucial, it acts as a minor player in the next two phases.

Data gathered from the case study conclude that the presence of stakeholders is essential to the evaluation process. So, opinions of customers and internal staffs are key factors that influence the opportunity of innovation as well as the possibility for NSD projects' success.

Through Figure 2.1, we bring evidence of how Bank X evaluates IT projects. We can see that the project evaluation process applied in practice includes some key activities mentioned by many authors.

Among traditional activities of the process of evaluation, respondents of the case study see setting indicators as one of the crucial activities of evaluation, emphasizing that referring back to the initial criteria helps to keep the project on the right track. Besides, most of the respondents mentioned collecting information through communication as a crucial part of evaluation process. In addition, the new model proposed in this book goes through the whole life cycle of the projects, providing a sufficient understanding of the evaluation process. Organizations are investing substantial funds into IT in an attempt to transform or re-engineer traditional business processes and ultimately improve productivity and profitability. Furthermore, considering that it takes a long time for investment in IT projects to show effects, it is desirable to actively adopt mid- and long-term impact evaluation.

In order to have more detailed perception on each of the following evaluation phases, the author would like to recommend the model illustrated in Figure 2.1.

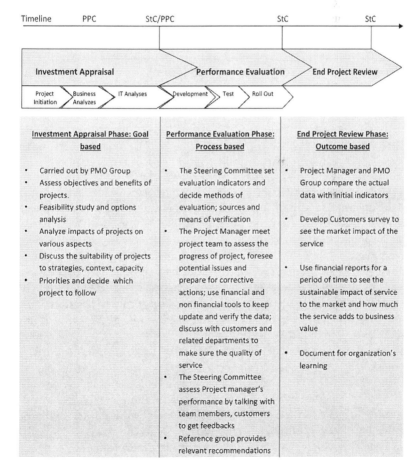

Figure 2.1 Evaluation process and purposes in IT projects

The investment appraisal phase

Investment appraisal is the first phase of the project life cycle. The initial project planning idea occurs during budgeting process for the upcoming year. This is important to better plan the costs and resources for the whole coming year. The Project Analysis serves to detail the project's cost/benefit, cost allocation, scope, structure, duration and schedule, thus presenting a full business case for the project.

The objective is to collect new ideas from business owners, consult and agree by direct Line Board Member, and sent to the Project Management Unit through a 'Screening' document, including the business case. This document represents the first high-level information prepared from initiators and sent to PMO at least five working days ahead of the nearest Project Portfolio Committee (PPC) meeting. The initiators are invited at the nearest PPC meeting to introduce their initiatives. The PPC decides whether these new initiatives should be further developed and evaluated or canceled. The PPC will nominate the Project Sponsor and Project Manager as well, based on the proposal of the PMO. All the decisions on approval/rejection of the initiatives presented during the PPC meeting as well as the nominated Sponsor and Project Manager are distributed to all participants through the meeting protocol.

Performance evaluation/control

This phase is divided into three parts: development, test and roll-out.

Development: The objective of Development phase is to create and develop the proposed and agreed solution. This phase's duration depends on the project solution complexity. Documents preparation that will support the project solution such as: regulation, operational procedures, manuals or directives, start at this phase but get finalized and approved only after a successful solution testing. The Project Manager and Project Officer follow every step of the development phase and report regularly per respecting reporting structures. Issues should be escalated within the project structure or, if needed, in PPC in real time.

The IT project manager has the responsibility to report and escalate immediately all issues during the development phase in time to the Project Manager, as well to take care of all required technical tests before moving on to functional testing. The IT project manager also must follow up with security to make sure a security analysis is done, controls are set and threat modeling (where applicable) is prepared.

Test: The testing phase consists of testing the functionality of the developed solution delivered by IT. Testing the developed solution and fixing all possible bugs that might come out during this phase is very important. This phase should be organized and monitored closely by the Project Manager/ Test Manager according to the approved test plan. The Project Manager, in collaboration with the Project Officer and Test Manager, prepares a detailed testing plan and delivers it to the responsible persons. It is crucial to have a well-defined and agreed-upon testing plan, based on predefined test cases. In addition, the Test Manager will take care to coordinate work on the following:

- Test environment
- Testing team
- Testing cycles/cases
- Communicating testing results and bugs fixed (report from bug repository system)
- Communicating testing progress and results to the Project Manager (if he/she not the same person) and to the IT manager

Roll-out: During this phase the new solution is deployed to the production environment. At the end of this phase, the solution in the production environment will be officially accepted by the users. Immediately after finalization of testing, the Project Manager has the responsibility to officially notify by email the Go Live date to the Project Sponsor, StC members and project core team members. The following are mandatory deliverables for the roll-out phase for most of the projects:

- ***Operational Procedures/Manual/Guidelines*** define the operational model including roles, processes and responsibilities for live implementation of the solution. This task should start in parallel with the development phase.
- ***Training module and training sessions*** will support the end users to have a good understanding of the new process/application/system.

After the project is rolled out, the **Stabilize** phase runs to ensure the new solution runs without any major issue in live production. It is the last phase of the project before handing the results.

End project review

The Project Manager ensures that the solution has been handed over to the responsible roles for its implementation in the live environment. The project is formally closed once the mandatory Project Closure documentation, or 'Project Completion Report', is signed off by the StC members, Project

Sponsor, Head of OPMD and Project Manager. The final signed document should be delivered to the PMO.

2.4 Challenges in evaluating IT projects

The data collected from our case study reveal several problems in evaluating IT projects. Some of them are related to the special characteristic of IT new services while some are associated with the project evaluation in general. Most of the challenges are in consistence with what have been found by previous studies on both project evaluation and new service development. The main difference between literature and practice is related to the tactical concern of academic versus the more dynamic attitude of practitioners. Although it remains a wide variety of reasons to justify investments in information systems (Irani and Love, 2001; Khalifa et al., 2000; Serafeimidis and Smithson, 2000) empirical evidence is needed to support the lack of widespread appraisal processes, financial or otherwise.

Main challenges faced by our respondents are described below.

Communication channel is a challenge in the evaluation process as respondents are having difficulties gathering reliable information. This is because sources of information are very important and lack of a good communication channel may lead to wrong evaluation results, which dramatically affect the decision-making procedure (McNamara, 1994). The problem calls for more attention and effort on how to enhance the feedback and feed-forward system within companies as well as with external stakeholders in practice.

Moreover, there should be a clear verification process to validate the information collected through the evaluation period, which improves the quality of the evaluation process. The willingness of people to participate and provide value comments allows the evaluation results to be more reliable which turn out to ensure the success of service projects.

It appears from the empirical study that *soft results* lead to evaluation difficulties or problems. The respondents brought to attention the difficulties associated with measuring intangible outcomes generated from the new service. This fact explained why there was a lot of effort in the academic work (Müller, 2003; Ballantine and Stray, 1998; Turner, 1995; Akalu, 2001) regarding the measurement techniques in the evaluation process such as NPV, IRR, payback period, discounted cash flow, etc.

The measurement issue becomes more serious in new service development projects due to the *intangibility characteristic* of service. Therefore, the evaluation of service projects is more complex than the evaluation of product projects. The opinion of the respondents in Bank X about the

Table 2.1 Challenges in evaluating IT projects

Source of the problem	Consequences
Communication channel	• Wrong results of evaluation • Low stakeholder involvement in the evaluation process
Soft results	• Difficulties in measurement of prestige, competitiveness, social effects
Intangibility characteristic	• Subjective evaluation
Customer involvement	• Subjective evaluation results • Time consuming
Risk assessment	• Difficulties in measurement of project profitability

difficulties of project evaluation was that in the case of IT evaluation there are hidden costs/benefits due to few specific or standard criteria to assess because of their intangible value.

A *customer involvement* issue was also raised by Thomas (1978) that customers find it risky to try the new service as they could not fully examine the quality of service prior to purchase. Therefore, evaluators in Bank X face difficulties to measure the usability of service, which is a crucial criterion of project impact evaluation.

The above characteristics of new service lead to difficulty in *risk assessment*. The respondents mentioned that it is difficult to measure risk, which is very high in service projects.

Table 2.1 summarizes the problems faced and illustrates the consequences in the evaluation process together with their impact.

2.5 Implications

Although the techniques applied are similar between academic and practices, the process of project evaluation for IT projects is rather complex due to the specific characteristic of service project. The choice of evaluation technique varies among sectors due to the particular feature of each (Gadrey et al., 1995; Lagsten, 2010). The consensus between research and practice on this issue implies that although setting indicators is a key activity in the evaluation process (Bellamy et al., 2001), it is more complex and needs to be applied in a more flexible way in practical world.

In particular, it requires more customer involvement and more discussion among different groups. This provides challenges for practitioners in measuring soft results generated by the development of the new service. It is also hard to get customers involved and obtain their valid comments as their answers are very emotional and inconsistent over the time. Therefore, the evaluation of NSD projects is usually considered as more subjective. More than that, in order to succeed, practitioners of service project evaluation ought to improve communication skill as it is needed to get appropriate information from different stakeholders.

3 Methods and techniques of evaluation

There is a multiplicity of evaluation methods and techniques associated with its own characteristics and purpose. Some of them can be used to help estimate the costs and benefits of the project and evaluate its worthwhileness. Others can be used only for estimation or only for evaluation. In practice, many companies combine parts of a number of methods and vary the methods to suit the situation. It is important to take into account that each method is applicable to a particular set of circumstances coupled with a number of factors influencing the success of a project. Therefore, it becomes crucial to match a method with the particular circumstances of the project. Evaluation context matters and contextual differences among specific cases are meaningful enough to undermine conclusions (Kaufman et al., 2014). Project performance should be rechecked using the same methods that were used for the initial reports (Radcliff, 2003). However, it is usually desirable to use two or more methods for analyzing a project. Different studies show that the project life cycle influences the evaluation process; therefore, it is reasonable to also classify methods and techniques of evaluation using the same logic, as we will discuss in this chapter. The author will analyze them in the context of IT projects and bring empirical evidence from real business life through our case study.

3.1 General evaluation methods and techniques throughout the project life cycle

Methods and techniques that are adopted in practice are not uniform throughout the process of evaluation. It depends on the phase of the project life cycle and the pertaining evaluation purpose. Accordingly, the author proposes a division of the process in specified stages and identified which methods and techniques value the most at each one.

Phase alignment is important to ensure optimal decision-making for programs and evaluation (Urban et al., 2014). But current methods for gathering data do not wholly capture program/project-related transformations, and grassroots ways of knowing yield legitimate data and can enrich programmatic efforts and evaluations (Dura et al., 2014). Let us discuss the evaluation methods and techniques for each phase of the project life cycle.

3.1.1 Investment appraisal: goal-based project evaluation

As we know, planning project evaluation is done at the very beginning of the project and prior to project implementation. It gives justification to choose projects among many others. At this phase, project evaluation provides information in terms of strategy such as rationale or justification and measuring the impact of the project.

Basically, the stages that deal with idea generation and screening, and qualitative and subjective criteria dominate the process. This is explained by the fact that there are not definite or consolidated concepts that can be conducive to qualitative analysis or to structured processes of evaluation, generally based on a quantitative approach. There is not yet a 'dominant design' or a well-defined concept, and as such there is no sure way to collect data or to rely on specific patterns of data. The technology that is being evaluated is presumably of a new vintage, or it possesses characteristics that may not be totally familiar to the evaluator, so there may not be recorded or familiar references on which a sound assessment can rely. Of course, this may vary with the incremental or radical nature of the technology. Qualitative assessments, experience and intuition become relevant factors in these stages. Later in this phase, as the product concept is consolidated and the requisites for data collection become much more established, quantitative assessments and financial methods/techniques take over, eventually dominating the later stages, whereby uncertainty is reduced and information retrieval is maximized.

Many authors consider this process similar to the investment decision-making as project actually is an investment. On one hand, the literature of project evaluation has placed a significant effort on providing various financial techniques that support the process. Some traditional methods used in project appraisal such as net present value (NPV), internal rate of return (IRR), payback period and discounted cash flow (DCF) were analyzed by many authors (Ballantine and Stray, 1998; Small, 1998; Müller, 2003). The role of these financial methods in evaluating projects is particularly used in the cost and benefit analysis. The tradeoff between accrued cost and future benefits related to all stakeholders should be carefully considered during the evaluation process. Investment appraisal techniques analyze whether to accept or reject the decision to undertake the project. Beside this, there are also some key investment factors that project managers should identify to ensure the financial gain of the projects (Müller, 2003). However, Akalu (2001) has criticized these methods on some of the problems below:

- 'The dis-conformity in the measurement techniques before and after the projects.
- The dynamic characteristic of project is ignored as changes during project implementation are not accommodated.
- Might forego some good projects as do not consider the intangible long-term benefits of projects.
- Do not consider the strategic importance of projects'

Another financial method as approach to investment evaluation, is real option, especially for R&D projects where value of management flexibility is crucial (Yeo, 2003). However, 'good analysis ties the details of strategy to financial implications' (Barwise et al., 1989, p. 85).

On the other hand, the intense employment of financial and quantitative side of projects defined it 'as myopic and misplaced' (Lopes and Flavell, 1998). Referring to theory on financial methods, combining with some non-financial aspects and investigation into some large British companies, Aldel-Kader and Dugdale (1998) emphasized the importance of intangible measurement criteria such as flexibility, delivery performance, quality improvement and reliability. Some strategic concerns during the evaluation process were identified by Greene (1988) and later by Alder (2000). The risk of failure would increase if other dimensions like the organizational and managerial aspects, political aspects, social acceptability, environmental problems, etc., depending on the nature of the project, are ignored. Not only financial and economic aspects but also other factors such as environmental impacts, employment effects, etc., should be taken into account to 'assist organization to decide whether a project concept is worth turning into reality' (Gardiner, 2005). Gardiner also stresses that the decision at this early stage before major resources have been consumed is critical and asking for careful evaluation and a feasibility study is a good tool for evaluating proposals. The competitive advantage, value chain, cost driver, strategic fit, relationship with stakeholders and uncertainty are critical to the success of projects that need to be addressed during the evaluation process. In addition, the consideration of the integration between the content, context and process and the inter-relatedness aspect allows management to be more flexible in making decisions (Stokdale and Standing, 2006).

In order to improve the success of a project, a simulation method is useful in testing how the project works and what the possible problems are if the project is launched (Doloi and Faafari, 2002). The need for a systematic evaluation of investment option, especially under the uncertain economy, has been emphasized by Mohamed and McCowan since 1999.

3.1.2 Performance evaluation: process-based project evaluation

The second phase of project evaluation, also called formative or ongoing evaluation, regards the progress and implementation evaluation. The purpose of evaluating projects in such a phase is to enable the company to decide whether it is worth going ahead or whether it is better to kill the project. In this phase, project evaluation provides information regarding operations and development, such as:

• Effectiveness in achieving expected outcomes
• Efficiency in optimizing resources
• Client satisfaction

During the second phase of the project life cycle, financial and development methods/techniques take over. We can especially distinguish development methods such as technical/system requirements, probability of project completion and user-friendly customer/supplier requirements as very important, as this phase is concerned with output (service-offered) indicators. These criteria are related to project activities. The marketing criteria, quality of information, and facilitating factors are what project evaluators look for in this phase in order to ensure the success of the new service. While it is obvious the marketing criteria targets at the goal of the project, the quality of information and facilitating factors show that the communication channel contributes significantly to the results of the evaluation of service projects.

The use of financial methods is mostly suggested to be employed in this phase as they offer adequate information through analyses such as cost-benefit analysis and ROI techniques, which provide important information to evaluate the ongoing progress of a project and make the right decisions. Also, APM Body of Knowledge (2006) suggests the use of investment appraisal techniques to provide a like-for-like comparison of options. According to APM, the evaluation on this phase of the project is considered as addition to ongoing monitoring and control process. International Development Research Centre (IDRC) discussed that this supports senior managers in decision-making. The evaluation process during product development is considered important to define the quality gate framework (Valeri and Rozenfeld, 2004). It could be said that ongoing evaluation is a key issue calling the attention of all persons involved in the project.

The performance evaluation phase is also important to project stakeholders (Greene, 1988). Crawford and Bryce (2003) share the same opinion, arguing that this type of evaluation, externally focused, is stakeholder-driven and emphasizes the effectiveness of the project, while monitoring is considered internally focused, management-driven and emphasizes on the efficiency of the project. However, the principal objective of evaluation in this phase is to ensure that implementation is on the 'right track' (Grabe, 1983). He argues that evaluation is seen as an opportunity for a direct contact with the project staff, dynamic situation in which activities are carried out as well as the possibility for curing malfunctions in project programming and implementation. Therefore, ongoing and ad hoc evaluation of programs and project in the course of implementation are called 'evaluation in vivo'. Evaluation during this phase of the project is carried on during a project to monitor activities to ensure they will achieve objectives and to alter the direction, redefine new objectives, modify approaches, etc., if necessary (Coutant and Cada, 1985; Steven et al., 1993). According to the literature, monitoring of ongoing work generally takes three forms, such as contacts between the agency project officer and the principal investigator, periodic progress or topical reports, and formal outside reviews.

According to Farbey et al. (1992), 'checks must be done to ensure that internal and external changes have not affected the feasibility of the project. At the same time progress on the project has to be assessed to ensure that the project is keeping within its budget'. They discuss that ROI techniques can be the natural choice, but there are projects, such as service development, which do not provide tangible benefits. Therefore, ROI is unable to capture many of the qualitative benefits that services such as IT bring (Farbey et al., 1992).

It is, however, necessary to take into consideration evaluation methods even though they can provide only information on the importance of alternative projects (Clemons, 1991). But Danks (1997) criticizes that it is vital for service projects to make clear decisions through clear evaluation methodologies. Despite the methods used during the performance evaluation phase, information provided is very important to the company and project stakeholders. 'Evaluation becomes a tool and base-data supplier for a rolling planning where goals and objectives are gradually advanced' (Grabe, 1983). In addition, ongoing evaluation occurring during this phase of evaluation produces information during the systems development process in order to help improve the product under development (Brown and Remenyi, 2002). According to PMBOK, the evaluation in this phase, considered also as monitoring and control, provides feedback in order to undertake actions that can correct or prevent deviations from project management plan.

3.1.3 End project review: outcome-based project evaluation

The third phase of project evaluation is done after project completion, to measure the outcomes of the project. Even though it is important to evaluate projects after their completion, not much literature can be found. Therefore, the study on this phase is limited to few authors. In this phase project evaluation provides information in terms of earning alternatives and lessons learned.

In this last phase, criteria of evaluation are considered impact indicators, which express the actual differences on the level of service offered after project implementation and the degree expected. The main methods/ techniques used are profitability and customer satisfaction. At a strategic level, evaluation performed in this last phase of the project is considered as important to generate information regarding the efficiency of the project by analyzing how successful the project has been in transforming the means through project activities into concrete project results.

This phase is also considered an 'ex-post' evaluation, where the main objective is 'to determine a starting point for further activities in the same field, to explore the relative cost, effectiveness and impact of alternative approaches, to identify common mistakes in comparable projects and to quantify such effects and impact patterns' (Grabe, 1983). It could be said that this evaluation can be considered a learning tool for the company such as improvement in productivity or career patterns in employment, 'snowballing' effects, etc. This idea is supported also by APM and PMI bodies of knowledge. It could be said that the evaluation after project completion makes an overall judgement about the effectiveness of a given project/program (Vakola, 2000; Scriven, 1967; Greene, 1988).

A post-project review or evaluation 'is a systematic inquiry concerning the merit of management and technical processes, and performance criteria. It helps identify root causes of success or failure and highlights improvement opportunities' (Huemann and Anbari, 2007; Anbari et al., 2008).

End-project review is also categorized into two phases: the 'testing phase' aimed at confirming effectiveness of final version and the 'routine phase' aimed at emphasizing quality assurance (Uhl, 2000). Despite this sub-division, evaluation after the project/program is valuable 'to mitigate poor project performance, demonstrate accountability and promote organizational learning for the benefit of future projects' (Crawford and Bryce, 2003). It ensures alignment of the performance measures with the project strategy.

There are two important objectives in the last phase of project evaluation: 'determining whether the contractor adequately carried out the goals and objectives of the work, as conceived in the proposal and contracted for in the work statement; determining whether the type of work done actually led to benefits to the resource commensurate with expectations and costs' (Coutant and Cada, 1985). The purpose is to determine if the participants' needs were met, if the problem was solved, if the project was efficient, if recipients of results were satisfied, what directions new programs might take, etc. Evaluation is considered important to measure the value of a project at the end (Chiesa and Masella, 1996). In contradiction to this, there are opinions that in practice the primary reason for outcome-based evaluations seems to be project closure and not project improvement (Kumar, 1990). According to Kumar, the reason is that post-implementation evaluations are being performed for the limited, short-term reason of formalizing the end of the development project, and they may not provide the more important long-term feedback-improvement benefits of the evaluation process.

However, it is important to convert lessons learned from the project evaluation into 'actions' (Morris and Pinto, 2010). It is not enough to go through the motions of conducting a lessons-learned exercise. When lessons are gathered and documented, there is a danger they will be put on a shelf where they remain unread and gather dust. For the lessons to be truly learned, they must be formulated in such a way that they ultimately lead to action. There are a number of approaches people take to promulgate learnings such as sharing lessons in informal meetings or maintaining a case study library of project experiences (Frame, 2002, in Morris and Pinto, 2010).

To summarize, Table 3.1 is presented as a brief overview of the evaluation methods and techniques during the project life cycle.

Table 3.1 Methods/techniques of project evaluation based on project life cycle

Type of project evaluation (PE)	*Investment appraisal: goal-based*	*Performance evaluation: process-based*	*End-project review: outcome-based*
Project life cycle	Conceptual phase	Ongoing phase	Close-out phase
Methods of evaluation	• Financial: NPV, IRR, payback period, DCF, real options • Non-financial: simulation, feasibility study	• Cost-benefit analysis • Investment appraisal techniques • Efficiency measurement	• Financial techniques (i.e., ROI) • Performance appraisal • Effectiveness measurement • End user satisfaction appraisal

(Continued)

Table 3.1 Continued

Type of project evaluation (PE)	Investment appraisal: goal-based	Performance evaluation: process-based	End-project review: outcome-based
Overall purpose	• Provide measures/ estimates to support investment decision-making process • Serve as baseline to set indicators for measuring success • Mostly used for appraising project proposals or selecting projects in portfolio management	• Improve overall performance of the project • Provide objective information • Provide measurements for control and decision-making process	• Make an overall judgement about the effectiveness of a given program • Provide lessons learned • Independent evaluation group is needed to avoid bias
Specific objective	• Narrow range of financial/ economic impact	• Wider range of human, organizational and economic impact	• Narrow range of financial/ economic impact
Timing	Before	During project implementation	After
Sources	Gardiner (2005), Müller (2003), Akalu (2001), Abdel-Kader and Dugdale (1998), Alder (2000)	Brown and Remenyi (2002); Grabe (1983); Caulley (1993); Banwell et al. (2003); Gudda (2011)	Vakola (2000); Scriven (1967); Segone (1998); Gudda (2011)

3.2 Evaluation methods and techniques in IT projects

Organizations evaluate their IT investments for several reasons, including to justify their investments, to enable organizations to decide between alternative projects, to control IT expenditures, to improve the investment selection process and to facilitate project management (Ballantine and Stray, 1998; Farbey et al., 1992; Ginzberg and Zmud, 1988). The evaluator needs to understand and describe the domain of the IS. The evaluator is not the domain expert, the stakeholders are. But the evaluator is responsible for depicting the IS-domain in a knowledgeable way (Lagsten, 2010).

Moreover, it is desirable to develop an evaluation method and guidelines that enable easy development of performance indicators for IT projects (Rosacker et al., 2008; Misso Kwon et al., 2010). It is also desirable to perform an ex-ante project evaluation and adopt a method that can help establish new projects or strategy and directions in addition to the model that verifies the effectiveness of the budget invested in IT projects.

Some financial measurements are suggested from the authors such as Small (1998), Ballantine and Stray (1998) and Müller (2003), which project evaluators could look at to assess the projects such as net present value (NPV), return on investment (ROI), discounted cash flow (DCF), internal rate of return (IRR) or payback period. ROI techniques can be the natural choice. But there are projects such as service development that do not provide tangible benefits. Therefore, ROI is unable to capture many of the qualitative benefits that new services such as IT bring (Farbey et al., 1992). There is an agreement in the literature that evaluations based on financial indicators are not appropriate to be considered in service projects and, in particular, IT proposals and IT investment performance (Parsons, 1983; Farbey et al., 1992; Hares and Royle, 1994; Remenyi, 1995; MIS, 1998 in Suwardy et al., 2003).

On the other hand, the use of non-financial techniques is also emphasized as necessary to evaluate the market impact of the new services. Simulation is a helpful method to test how the project works and what are the possible problems of the project (Fox and Baker, 1985; Doloi and Jaafari, 2002).

There are some critics that it is vital for IT projects to take clear decisions through clear evaluation methodologies (Danks, 1997). Despite the method used during the evaluation of a project, information provided is very important to the company and project stakeholders. The importance of ongoing evaluator/client communication as a tool which serves to facilitate transfer of research findings (Honadle et al., 2014). Moreover, developmental evaluation provides insights into how interdisciplinary teams may need to change course and conduct a valuable evaluation.

Since 1991, Clemons identified that some form of evaluation is needed even if it is only to provide a ranking of alternative projects. Later, Clemons and Demombynes (2011) and Féris et al. (2017) show that estimates of the projects' effects depend heavily on the evaluation method.

Given the intangibility trait of services in IT projects, it becomes crucial to apply evaluation methods able to capture both quantitative and qualitative traits.

This book provides insights to identify which are the most appropriate methods/techniques for evaluating IT projects.

Two broad categories of IT project evaluation techniques commonly used in evaluating IT investment opportunities can be identified: financial and qualitative (Bannister and Remenyi, 2000). This may be because while the selection process is recognized as important, information systems projects involve group-oriented activities subject to the benefits and problems of group dynamics, interactions, coordination and communication (Omitaomu and Badiru, 2007). Thus, qualitative approaches often override quantitative approaches, which may require more detailed data than are reasonably available.

A survey conducted by Bacon in 1992 identified three groups of important approaches to be used in IST project evaluation (Investment Decisions) classified as financial approach, management approach and development approach. Later, the criteria perceived as important by 88 IS managers in the context of strategic relevance were explored by Jiang and Klein (1999). They found three groups of evaluation criteria, with financial criteria and organizational needs grouped together, technical IS matters grouped with user need and management support and factors such as customer/supplier requirements, industry standards, legal requirements and response to competition as the third group.

The book provides empirical evidence from the case study conducted in Bank X, in order to give a practical insight to this problem.

3.2.1 Empirical evidence of the evaluation methods and techniques

Based on the empirical data gathered from the case study and cross-checked through interviews, the main project evaluation methods and techniques used to evaluate IT projects fall under the qualitative group, given the intangible nature of IT as a service project.

Appointed departments list their project needs/requirements. Then, the information gathered from the project evaluation serves to several groups and is used in different stages of the project life cycle:

1   In the ***investment appraisal phase***, board members make the decision among several projects of which is more valuable based on the evaluation results. At the *project start-up* moment, project evaluation is performed in order to analyze the cost/benefit of the project. This is done through:

•   *Business case.* This analysis is important because the project could generate negative payoff.

- *Technical analyses* (this evaluation is done by the IT team). The release manager draws up a prototype IT solution proposal as well as the costs schedule for each proposed solution.
- *Business case vs. technical analyses*. The evaluation team compares the business case with the technical proposed solution. This happens in case actual costs are 20% more than planned costs, and the board members have to make a decision. If actual costs are ≤20% more than the planned costs, the proposed solution is presented to the StC to be selected as the final solution.
- *Project charter* is a detailed preview of cost/benefit analyses and serves as a working plan.
- *Stakeholders analyses* to evaluate their attitudes and participation willing/impact.

2   At the ***performance evaluation*** phase, the StC is interested to know how the project is going. Each project is under the responsibility of a project manager and his project officer of PMO. Another person from the PMO controls the project in parallel with regards to respecting the methodology, quality of deliverables, scope, timing and project budget tracking. Based on the project ID defined at the beginning of each project, all project costs reported are aligned with Cost and Budget Monitoring unit as well as with the payments done via i-procurement. Through the Cost and Budget Monitoring unit it is also assured that the correct figures are reported to Finance and reflected in relevant applications. This is due to Budget, Actual Cost and Forecast. Most current picture of overall project portfolio (scope, timing, cost) is reported regularly (approximately every six weeks) to the PPC/Management Board.

The financial aspect of a project is essential to evaluating projects since it provides the project sponsor or evaluators a numerical idea of how much the project will cost, how much the company will win and for how long the revenue could cover the amount of money invested or how much shareholders gain from the project. These are visible information that project evaluators could at least measure and compare over time. However, these numbers have to be based on realistic data and use suitable projection standards to avoid over- or under-estimation. During project implementation, the data have to be updated to see if the project is running on-budget or how much changes are influencing shareholders' value.

Most respondents confirmed that those basic financial indicators are used all the time when they do evaluation for IT projects. On the other hand, the use of non-financial techniques is also emphasized as necessary to evaluate the market impact of the new service. The following methods are being used to evaluate the performance of the project:

- During the *development phase*, user live acceptance technique and operational procedures take place.
- *Business requirement specifications* to define the required target to be achieved through the project from the business point of view, what is critical and should be improved (process, product, service or system) in order to fulfill the business needs. In addition, the respondents mentioned one of the three main criteria that have been strongly recommended by the triangle of PMI model is *scope*.
- *Simulation*: Bank X during the *test case* uses concrete data. Moreover, it uses *test scenarios* to identify requirements. Simulation is a helpful development method to test how the project works and determine possible problems (Fox and Baker, 1985; Doloi and Jaafari, 2002). The respondents from the empirical study also emphasize that simulation, survey or direct communication are even more important techniques to customize their service.
- Back tracker is a tool accessed by all test users to generate a final report called a *users acceptance test* (UAT). The UAT also mentioned by Morris and Pinto (2010) is considered an important evaluation method because it is undertaken by the players for whom the deliverable is being developed. Customer satisfaction is very important to measure quality, which is more sensitive to service projects (Tukel and Walter, 2001), and defines how successful a project is in terms of market impact.
- *Piloting:* For some projects (e.g., mobile banking), the service development turns live, but this happens first at one branch or for bank staff only.
- *Meeting minutes* keep track of the project and determine next steps.

3 ***End-project review***. This phase is well known for the *lessons learned* during the project implementation. The project manager is responsible for drafting lessons learned.

- The *project completion report* is signed by the project manager and the StC members. It compares the realized performance with expected performance and serves to draw the final report of project closure.

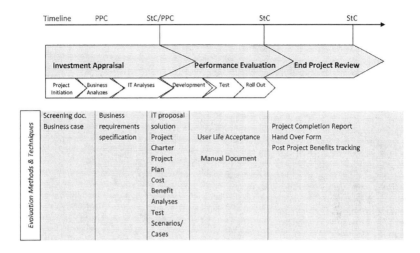

Figure 3.1 Evaluation methods and techniques in IT projects

Figure 3.1 illustrates evaluation techniques used throughout the project life cycle.

3.3 The evaluation framework

The author has come up with an evaluation framework (see Figure 3.2) by combining both findings from conducting a systematic literature review on the topic and findings from the empirical study. This is the main contribution of the book, intended to guide practitioners on how to successfully evaluate their IT projects.

The proposed framework illustrates how to evaluate service projects, especially IT projects, and choose the appropriate evaluation methods and techniques depending on the following parameters:

- The purpose of evaluation – reasoning evaluation of the projects and pointing out the evaluation benefits.
- The phase of the project life cycle – helps to realize how to evaluate the project and decide what the appropriate methods/techniques are.
- The context relevance/approach (management, financial and development) – depending on the decision-making type.
- The nature (quantitative and qualitative).

How to evaluate IT projects	Evaluation process		
	Investment appraisal	Performance evaluation	End project review
Evaluation methods and techniques			
WHAT to evaluate **(purpose of evaluation):**	Goal-based	Process-based	Outcome-based
WHY to evaluate: **(in terms of decision-making):**	• Strategic • Financial	• Operational • Financial	• Strategic • Financial
HOW to evaluate **(the methods and techniques of evaluation):**	• Financial: ratio (payback period, ROI, CBA, etc.), discounting (NPV, IRR, etc.), future (real options, etc.) • Non-financial: simulation, feasibility study, stakeholder analyses	• Cost-benefit analysis • Investment appraisal techniques • Efficiency measurement	• Financial techniques (i.e, ROI) • Performance appraisal • Effectiveness measurement • End-user satisfaction appraisal
Evaluation outcomes:	*Impact* Analyses of the overall effects of the project, the contribution of the project purpose to the overall objectives.	*Efficiency* Analyses on how successful the project has been in transforming the means (e.g., the resources and inputs allocated to the project) through project activities into concrete project results. Provides stakeholders with information on inputs/costs per unit produced.	*Effectiveness* Analyses on how well the production of project results contributes to the achievement of the project purpose and objectives.

Figure 3.2 Suggested framework for project evaluation

3.4 Implications

The author's focus on the processes and methods/techniques coupled with the difficulties faced suggest that the evaluation of projects is even more demanding than is suggested in practice.

The case study conducted at Bank X in Albania proved the importance of project evaluation throughout the project life cycle and the application of adequate methods and techniques at each phase. These applications provided an important incentive for increasing project success and improving decision-making through appropriate methods and techniques based also at the purpose of the evaluation.

Appendix 1

Interview guide

The evaluation process could be applied prior to, during or after projects, depending on your company's practice and purposes.

Introduction key components: • Purpose • Confidentiality • Duration • How interview will be conducted • Opportunity for questions • Signature of consent	I want to thank you for taking the time to meet with me today. My name is ***Eriona Shtëmbari*** and I would like to talk to you about your experiences participating in the project evaluation. As one of the components of my overall research study on project evaluation, I am interested in the evaluation process for IT Projects. I am particularly looking at the evaluation methods/techniques you are using during the project life cycle. These can be used either at the decision-making process to select new projects, during the project execution to track its progress, or at project completion in order to capture lessons that can be used in future projects. The interview should take less than an hour. I will be taping the session because I don't want to miss any of your comments. Although I will be taking some notes during the session, I can't possibly write fast enough to get it all down. Because we're on tape, please be sure to speak up so that we don't miss your comments. All responses will be kept confidential. This means that your interview responses will not be shared and I will ensure that any information included in the report does not identify you as the respondent. Please consider that you don't have to talk about anything you don't want to and you may end the interview at any time. The evaluation process could be applied prior to the projects, during or after projects depending on your company's practice and purposes. Are there any questions about what I have just explained? Are you willing to participate in this interview?

Interviewee Date

Questions	*General Questions*
	• What are your experiences and/or roles with project evaluation, especially with service projects?
	• How does your company form project/evaluation teams? What is the relationship between them?
	• What was the level of managerial commitment to IT project evaluation?
	• How much are stakeholders (end users, vendors, sponsors, etc.) involved/influence the evaluation process?
	Project evaluation in IT projects
	• What are the purposes of the evaluation process in your company?
	• Could you please tell me about *the process of evaluation for IT projects* applied?
	• What are the *techniques/methods* that you use for evaluating IT projects?
	• How does the bank evaluate the cost-benefit of its IT environment?
	• How do you decide to invest in the new service?
	• How do you ensure the IT project is on the right track?
	• Based on what can you say the project is a success?
	Final questions
	• What are the difficulties and challenges in the evaluation process?
	• Could you recommend an evaluation method that you use for your work?
	• Does your company have any official framework for evaluating IT projects? Is it possible to access it?
Closing key components	Is there anything more you would like to add?
	I'll be analyzing the information you and others gave me and prepare a draft report shortly. I'll be happy to send you a copy to review at that time, if you are interested.
	Thank you very much for your kind attendance and time!

Appendix 2

Case study abbreviation list

Acronyms	Descriptions
BRS	Business Requirements Specifications
CBA	Cost Benefit Analysis
MD	Man Days
OPMD	Organization and Project Management Department
PM	Project Manager
PMO	Project Management Office
PO	Project Officer
PPC	Project Portfolio Committee
StC	Steering Committee
UAT	User Acceptance Test

Appendix 3
Main literature

Project evaluation	IT	Methods & techniques of evaluation
The APMBoK and the PMBOK® Guide – *standard authorized guidelines* that support the project management domain. The APMBoK section on evaluation and control, which includes many of the traditional tools associated with project evaluation, emphasizes the importance of project evaluation and control *during the project life cycle*. This is an important difference between the *APMBoK* and the *PMBOK® Guide* Gudda (2011) argues that evaluation maybe categorized in two kinds of stages – formative evaluation and summative evaluation. The purpose of formative evaluation is to assess whether the project is being conducted as planned, while the summative evaluation the purpose is to assess a mature project's success in reaching its stated goals.	Stewart (2007): It is obvious that IT is becoming very important and is recently implemented in many organizations for strategic use. Gunasekarana et al. (2001): Adoption of IT initiatives can be more lengthy, expensive and complex than other service projects. Wu and Ong (2008): IT investments also tend to have a high failure rate that might have potentially devastating impacts. Nelson (2007): According to IT project management statistics, two out of every three IT projects fail partially or totally. Müller (2003); Ballantine and Stray (1998); Turner (1995); Akalu (2001) regarding the *soft results* measurement using	Kaufman et al (2014) argue that evaluation context matters and that contextual differences among specific cases are meaningful enough to undermine conclusions. Thomas (1978) states that the key to successful project evaluation is to decide on evaluation methods and techniques and determine the value that evaluation brings and what the stakeholders desire. Radcliff (2003): Project performance should be rechecked using the same methods that were used for the initial reports. Bacon (1992) and Irani et al., 1995, in Syewart (2008): Strategic approach for classification of evaluation methods and techniques in IT.

(Continued)

Continued

Project evaluation	IT	Methods & techniques of evaluation
Mertens and Wilson (2012), Gramham (2006) consider *project evaluation process as crucial for the success of projects.*	techniques in evaluation process such as NPV, IRR, payback period, discounted cash flow, etc.	
McNamara (1994) stresses the importance of *sources of information*, the reliability and validity of feedbacks that contribute to decision-making in program evaluation.	Farbey et al. (1992) proposed a matrix method for matching an IT project with a suitable evaluation technique. This matching method has been developed *ex-post* from the investigation of 16 case studies, but it has not been validated.	
Steven et al. (1993) and Örtengren (2004) present evaluation activities in a linear order.		
Bellamy et al. (2001) suggests an *integrated evaluation process* that takes into account the correlation among different activities to ensure the success of projects, and continuously generated knowledge for the next projects.		

References

Abdel-Kader, M. G. and Dugdale, D. (1998). Investment in advanced manufacturing technology: A study of practice in large U.K. Companies. *Management Accounting Research*, 9 (3), 261–284.

Adler, R. W. (2000). Strategic investment decision appraisal techniques: The old and the new. *Business Horizons*, 43 (6), 15–22.

Ahlemann, F., Arbi, F. E., Kaiser, M. G. and Heck, A. (2013). A process framework for theoretically grounded prescriptive research in the project management field. *International Journal of Project Management*, 31, 43–56.

Akalu, M. M. (2001). Re-examining project appraisal and control: Developing a focus on wealth creation. *International Journal of Project Management*, 19 (7), 375–383.

Akalu, M. M. (2003). The process of investment appraisal: The experience of 10 large British and Dutch companies. *International Journal of Project Management*, 21 (5), 355–362.

Alam, I. and Perry, C. (2002). A customer-oriented new service development process. *Journal of Service Marketing*, 16 (6), 515–534.

Anbari, F. T. et al. (2008). Post-project reviews as a key project management competence. *Technovation*, 28 (10), 633–643.

Andersen, E. S. et al. (2002). Evaluation of Chinese projects and comparison with Norwegian projects. *International Journal of Project Management*, 20 (8), 601–609.

APM. (2006). *APM Body of Knowledge*. Buckinghamshire: Ingmar Folkmans.

Avlonitis, G. J., et al. (2001). An empirically-based typology of product innovativeness for new financial services: Success and failure scenarios. *The Journal of Product Innovation Management*, 18 (5), 324–342.

Bacon, J. (1992). The use of decision criteria in selecting information systems/technology. *Investments MIS Quarterly*. 16 (3), 335–353.

Ballantine, J. and Stray, S. (1998). Financial appraisal and the IS/IT investment decision-making process. *Journal of Information Technology*, 13 (1), 3–14.

Bannister and Remenyi (2000). Acts of faith: Instict, Value and IT investment decisions. *Journal of Information Technology*, 15 (3), 231–241.

Banwell, L. et al. (2003). Evaluation, impact and outcomes – the Jubilee project. *Performance Measurement and Metrics*, 4 (2), 79–86.

Barclay, C. (2008). The Project Objectives Measurement Model (POMM): An alternative view to information systems project measurement. *Electronic Journal of Information Systems Evaluation*, 11 (3), 139–153.

Bardhan, et al. (2004). Prioritizing a Portfolio of Information Technology Investment Projects. *Journal of Management Information Systems*, 21 (2).

Barwise, P. et al. (1989). Must finance and strategy clash? *Harvard Business Review*, 67 (5), 85–90.

Baxter, P. and Jack, S. (2008). Qualitative case study methodology: Study design and implementation for novice authors. McMaster University, West Hamilton, Ontario, Canada. *The Qualitative Report*, 13 (4), 544–559.

Bellamy, J. et al. (2001). A systems approach to the evaluation of natural resource management initiatives. *Journal of Environmental Management*, 63 (4), 407–423.

Berry, L. (1980). Services marketing is different. *The Journal of Business*, 30 (3), 24–29.

Bowers, M. (1989). Developing new services: Improving the process makes it better. *Journal of Services Marketing*, 3 (1), 15–20.

Brandon, et. al (2014). Issues of rigor and feasibility when observing the quality of program implementation: A case study. *Evaluation and Program Planning*, 44, June 2014, 75–80.

Brown, A. and Remenyi, D. (2002). *Ninth European Conference on Information Technology Evaluation*. Reading, UK: MCIL.

Bryman, A. and Bell, E. (2003). *Business Research Methods*. New York: Oxford University Press.

Caulley, D. (1993). Evaluation: Does it make a difference? *Evaluation Journal of Australia*, 5 (2), 3–15.

Chapman, C., Ward, S. and Harwood, I. (2006). Minimising the effects of dysfunctional corporate culture in estimation and evaluation processes: A constructively simple approach. *International Journal of Project Management*, 24 (2), 106–115.

Chapman, R. L. and Soosay, C. (2003). Innovation in logistic services and the new business model A conceptual framework. *International Journal of Physical Distribution & Logistics Management*, 33 (7), 630–650.

Chiesa, V. and Masella, C. (1996). Searching for an effective measure of R&D performance. *Management Decision*, 34 (7), 49–57.

Clemens, M. and Demombynes, G. (2011). When does rigorous impact evaluation make a difference? The case of the millennium villages. *Journal of Development Effectiveness*, 3 (3), 305–339, 35. doi:10.1080/19439342.2011.587017

Clemons, E. K. (1991). Making the investment decision: Evaluating strategic opportunities in information technology. *Communications of the ACM*, 34 (1), 22–36.

Cooper, D. R. and Schindler, P. S. (2008). *Business Research Methods: Second European Edition*. McGraw-Hill Higher Education; INTSTDT edition.

Coutant, C. C. and Cada, G. F. (1985). *Analysis and Development of a Project Evaluation Process*. Portland, OR: U.S. Department of Energy, Bonneville Power Administration, Division of Fish and Wildlife.

Cowell, D. W. (1988). New service development. *Journal of Marketing Management*, 3 (3), 296–312.

Crawford, P. and Bryce, P. (2003). Project monitoring and evaluation: A method for enhancing the efficiency and effectiveness of aid project implementation. *International Journal of Project Management*, 21 (5), 363–373.

Creswell, J. W. (2003). *Research Design: Qualitative, Quantitative, and Mixed Method Approaches*. Second edition. Thousand Oaks, CA: Sage.

Danks, D. (1997). When big means ugly. *MIS*, July, 56–60.

Day, G. S. (2007). Is it real? Can we win? Is it worth doing? *Harvard Business Review*, 85 (12), 110–120.

De Brentani, U. D. (1991). Success factors in developing new business services. *European Journal of Marketing*, 25 (2), 33–57.

De Brentani, U. D. (1995). New industrial service development: Scenarios for success and failure. *Journal of Business Research*, 32 (2), 93–103.

De Falco and Macchiaroli (1998). Timing of control activities in project planning. *International Journal of Project Management*, 16 (1).

De Reyck, et al. (2005). The impact of project portfolio management on Information Technology Projects. *International Journal of Project Management*, 23, 524–537.

Denzin, N. K. and Lincoln, Y. S. (1994). *Handbook of Qualitative Research*. Thousand Oaks, CA: Sage.

Dolfsma, W. (2004). The process of new service development – Issues of formalitization and appropriability. *International Journal of of Innovation Management*, 8 (3), 319–317.

Doloi, H. and Jaafari, A. (2002). Conceptual simulation model for strategic decision evaluation in project management. *Logistics Information Management*, 15 (2), 88–104.

Dos Santos, B. L. and Sussman, L. (2000). Improving the return on IT investment: The productivity paradox. *International Journal of Information Management*, 20, 429–440.

Dover, P. A. (1987). Innovation in banking: The in-home computerised banking example. *International Journal of Bank Marketing*, 5 (1), 39–54.

Dura et al. (2014). What counts? For whom?: Cultural beacons and unexpected areas of programmatic impact. *Evaluation and Program Planning*, 44, 98–109.

Edgett, S. and Jones, S. (1991). New product development in the financial service industry: A case study. *Journal of Marketing Management*, 7 (3), 271–284.

Eisenhardt, K. M. and Graebner, M. E. (2007). Theory building from cases: Opportunities and challenges. *Academy of Management Journal*, 50 (1), 25–32.

Erlandson, D. A., Harris, E. L., Skipper, B. L. and Allen, S. D. (1993). *Doing Naturalistic Inquiry: A Guide to Methods*. Newbury Park, CA: Sage.

Farbey, B., Land, F. and Targett, D. (1992). Evaluating investments in IT. *Journal of Information Technology*, 7 (2), 109–122.

Féris, M.A.A., Zwikael, O. and Gregor, S. (2017). QPLAN: Decision support for evaluating planning quality in software development projects. *Decision Support Systems. Elsevier*, 96, 92–102.

Fox, G. E. and Baker, N. R. (1985). Project selection decision-making linked to a dynamic environment. *Management Science*, 31 (10), 1272–1285.

Frechtling, J. (2002). *The 2002 User Friendly Handbook for Project Evaluation*. Arlington: National Science Foundation.

Gadrey, J., Gallouj, F. and Weinstein, O. (1995). New modes of innovation How services benefit industry. *International Journal of Service Industry Management*, 6 (3), 4–16.

Gallouj, F. and Weinstein, O. (1997). Innovation in services. *Research Policy*, 26 (4–5), 537–556.

Gardiner, P. (2005). *Project Management: A Strategic Planning Approach*. New York: Palgrave MacMillan.

Ghapanchi, A. H. et al. (2012). A methodology for selecting portfolios of projects with interactions and under uncertainty. *International Journal of Project Management*, 30 (7), 791–803.

Gibbert, M. et al. (2008). Research notes and commentaries. What passes as a rigorous case study? *Strategic Management Journal*, 29, 1465–1474.

Gifford, S. and Wilson, C. (1995). A model of project evaluation with limited attention. *Economic Theory*, 5 (1), 67–78.

Globerson and Zwikael (2002). The impact of the project manager on project management planning processes. *Project management journal*, 33 (3), 58-64.

Ginsberg, M. J. and Zmud, R. W. (1988). Evolving criteria for *Information Systems assessment. Information Systems Assessment: Issues and Challenges: proceedings of the IFIP WG 8.2 Working Conference on Information Systems Assessment: Issues & Challenges*, N. Bjorn-Andersen and G. B. Davis (Eds.), North Holland, Amsterdam, 41–55.

Grabe, S. (1983). *Evaluation Manual*. UNESCO, Imprimerie de la Manutention, Mayenne, France.

Graham, D. (2006). Evaluation Lessons: Green groups must measure gains. *Regeneration & Renewal*, Haymarket Business Publications Ltd, London.

Greene, J. G. (1988). Stakeholder participation and utilization in program evaluation. *Evaluation Review*, April, 12 (2), 91–116.

Griffin, A. (1997). PDMA research on new product development practices: Updating trends and benchmarking best practices. *Journal of Product Innovation Management*, 14 (6), 429–458.

Grönroos, C. (1990). *Service Management and Marketing: Managing the Moments of Truth in Service Competition*. Lexington, MA: Lexington Books.

Grönroos, R. (1998). *Service Marketing Theory: Back to Basics*. Working paper. CERS, Finland.

Grönroos, R. (2000). *Service Management and Marketing: A Customer Relationship Management Approach*. Second edition. England: John Wiley and Sons.

Gunasekarana, et al. (2001). Performance measures and metrics in a supply chain environment. *International Journal of Operations & Production Management*, 21 (1–2), 71–87.

Gudda (2011). A Guide to Project Monitoring & Evaluation, AUTHORHOUSE, US.

Hares, J. and Royle, D. (1994). *Measuring the Value of Information Technology*. Chichester: John Wiley & Sons.

Heany, D. F. (1983). Degrees of product innovation. *The Journal of Business Strategy*, 3 (4), 3.

Hertog, P. (2000). Knowledge-intensive business services as co-producers of innovation. *International Journal of Innovation Management*, December, 4 (4), 491.

Highsmith (2004). Agile Project Management: Creating Innovative Products. Addison-Wesley Professional.

Honadle, et al. (2014). Developmental evaluation and the 'Stronger Economies Together' initiative in the United States. *Evaluation and Program Planning*, 43, April 2014, 64–72.

Huemann, M. and Anbari, F. (2007). Project auditing: A tool for compliance, governance, empowerment, and improvement. *Journal of Academy of Business and Economics*, 7 (2), 9–17.

Huff, A. S. (1999). *Writing for Scholarly Publication*. Thousand Oaks, CA: Sage.

Huff, A. S. (2009). *Designing Research for Publication*. Thousand Oaks, CA: Sage.

Huff, A. S. and. Möslein, K. M. (2009). Framing research on service. In Donald, D. B. and Ketchen, D. J. (Eds.) *Research Methodology in Strategy and Management*, Volume 5, Emerald Group Publishing Limited, 179–212.

International Development Research Centre. (IDRC). Canada. www.idrc.ca

Irani, et al. (2002). Applying concepts of fuzzy cognitive mapping to model: The IT/ IS investment evaluation process. *International Journal of Production Economics*, 75 (1–2), 199–211.

Irani Zand Love P.E.D. (2001). The propagation of technology management taxonomies for evaluating investments in information systems. *Journal of Management Information System*, 17(3), 161–177.

Jackson, B. (2000). Designing projects and project evaluations using the logical framework approach. *IUCN Monitoring and Evaluation (M&E) Initiative*, 1–11. www. infra.kth.se/courses/1H1146/Files/logicalframeworkapproach.pdf (Accessed online 29 October 2008).

Jackson, R. W. and Cooper, P. D. (1988). Unique aspects of marketing industrial services. *Industrial Marketing Management*, 17 (2), 111–118.

Jiang and Klein (1999). Project selection criteria by strategic orientation. *Information & Management*, 36, 63–75.

Johne, A. and Storey, E. C. (1998). New service development: A review of the literature and annotated bibliography. *European Journal of Marketing*, 32 (3), 184–251.

Johnson, S. P., Menor, L. J., Roth, A. V. and Chase, R. B. (2000). *A Critical Evaluation of the New Service Development Process: Integrating Service Innovation and Service Design*. Thousand Oaks, CA: Sage, 1–32.

Jong, J. D., Bruins, A., Dolfsma, W. and Meijaard, J. (2003). Innovation in service firms explored: What, how and why? Literature review. *Business and Policy Research*. www.ondernemerschap.nl/pdf-ez/B200205.pdf (Accessed online 1 November 2008).

Jong, J. and Vermeulen, P. (2003). Organising successful new service development: A literatue review. *Management Decision Journal*, 41 (9), 844–855.

Keil, M. et al. (2002). Reconciling user and project manager perceptions of IT project risk: A Delphi study. *Information Systems Journal*, 12, 103–119.

Kelly, D. (2000). New service development: Initiation strategies. *International Journal of Service Industry Management*, 11 (1), 45–62.

Kaufman, et. al (2014). Evaluating participatory decision processes: Which methods inform reflective practice?, *Evaluation and Program Planning*, 42, February 2014, 11–20.

Khalifa, K. (2000). Building Strong Management and Responding to Change. Banking Institutions in Developing Markets, 1.

Klapka, J. and Pinos, P. (2002). Decision support system for multi criteria R&D and information systems projects selection. *European Journal of Operation Research*, 140 (2), 434–446.

Klivans, J. M. (1990). Launching a financial service: A case study in persistence. *The Journal of Business Strategy*, 11 (5) 8–11.

Kumar, K. (1990). Post implementation evaluation of computer-based information-systems: Current practices. *Communications of the ACM*, 33 (2). www.portal. acm.org (Accessed online 24 October 2008).

Lagsten (2010). Stakeholder based method for evaluating Information Systems – The VISU method. Örebro universitet, Handelshögskolan vid Örebro universitet.

Langeard, E. and Eiglier, P. (1983). Strategic management of service development. In Berry, L. L. et al. (Eds.) *Emerging Perspectives on Services*. Chicago, IL: AMA, 68–72.

Levitt, T. (1976). The industrialization of service. *Harvard Business Review*, 54 (5), 63–74.

Liang, W-Y. (2003). The analytic hierarchy process in project evaluation: An R&D case study in Taiwan. *Benchmarking: An International Journal*, 10 (5), 445–456.

Lopes, M. D. and Flavell, R. (1998). Project appraisal a framework to assess non-financial aspects of projects during the project life cycle. *International Journal of Project Management*, 16 (4), 223–233.

Lovelock, C. (1984). *Developing and Implementing New Services*. Chicago: American Marketing Association.

Lovelock, C. H. (1983). Classifying services to gain strategic marketing insights. *Journal of Marketing*, 47 (3), 9–20.

Maister, D. H. and Lovelock, C. H. (1982). Managing facilitator services. *Sloan Management Review*, Summer, 23, 19–31.

McAvoy, J. (2006). Evaluating the evaluations: Preconceptions of project post-mortems. *The Electronic Journal Information Systems Evaluation*, 9 (2), 65–72.

McConnell, S. (1996). *Rapid Development: Taming Wild Software Schedules*. Microsoft Press. ISBN-10: 1556159005. ISBN-13: 978–1556159008.

McNamara, J. F. (1994). *Surveys and Experiments in Education Research*. Lancaster, PA: Technomic.

Menor, L. J. and Roth, A. V. (2007). New service development competence in retail banking: Construct development and measurement validation. *Journal of Operations Management*, 25 (4), 825–846.

Menor, L. J., Tatikonda, M. V. and Sampson, S. E. (2002). New service development: Areas for exploitation and exploration. *Journal of Operations Management*, 20 (2), 135–157.

Mertens and Wilson (2012). Program Evaluation Theory and Practice: A Comprehensive Guide By Guilford Press.

Messner, J. I. and Sanvido, V. E. (2001). An information model for project evaluation. *Engineering, Construction and Architectural Management*, 8 (5–6), 393–402.

Meyer, M. H. and De Tore, A. (2001). Perspective: Creating a platform-based approach for developing new services. *Journal of Product Innovation Management*, 18 (3), 188–204.

Milis, K. and Mercken, R. (2004). The use of the balanced scorecard for the evaluation of Information and Communication Technology projects. *International Journal of Project Management*, 22 (2), 87–97.

Misso Kwon, et al. (2010). Self-evaluation system of IT projects in Korean central government: Instituion and practices. *Proceedings of the 4th European Conference on Information Management and Evaluation: Universidade Nova de Lisboa*, Lisbon, Portugal.

Mohamed, S. and McCowan, A. K. (2001). Modelling project investment decision under uncertainty using possibility theory. *International Journal of Project Management*, 19 (4), 231–241.

Morris and Pinto (2010). *The Wiley Guide to Project Organization and Project Management Competencies*, John Wiley & Sons.

Müller, R. (2003). Commercial aspects of project management, an introduction. *Umeå School of Business and Administration, Umeå University*, 1–14.

Nelson, R. R. (2006). Tracks in the snow: IT projects are usually judged successes or failures when they go live. But look back and you will see the real judgment comes later, and that requires a new set of value criteria. *CIO Framingham*, 19 (22), 1.

Nelson, R. R. (2007). IT project management: Infamous failures, classic mistakes and best practices. *MIS Quarterly Executive*, 6 (2), 67–78.

Newbert, S. L. (2007). Empirical research on the resource based view of the firm: An assessment and suggestions based for future research. *Strategic Management Journal*, 28 (2), 121–146.

Nolan, R. and McFarlan, F. W. (2005). Information technology and the board of directors. *Harvard Business Review*, 83, 96–106.

Nurmi, A. et al. (2011). When does rigorous impact evaluation make a difference? The case of the Millennium Villages. *Business Process Management Journal*, 17 (5), 711–731, 21. doi:10.1108/14637151111166150

OECD. (2000). *Promoting Innovation and Growth in Services*. Paris: Organisation for Economic Cooperation and Development.

Oral, M., Kettani, O. and Lang, P. (1991). A methodology for collective evaluation and selection of industrial R&D projects. *Management Science*, 37 (7), 871–885.

Örtengren, K. (2004). A summary of the theory behind the LFA method: The logical framework approach. *Sida*, 1–40.

Osman, K. et al. (2005). Multi-attribute information technology project selection using fuzzy axiomatic design. *Journal of Enterprise Information Management*, 18 (3), 275–288.

Parsons, G. L. (1983). Information technology: A new competitive weapon. *Sloan Management Review*, 25 (1), 3–14.

Pena-Mora, et al. (1999). Interaction Dynamics in Collaborative Civil Engineering Design Discourse: Applications in Computer Mediated Communication, *Journal of Computer-Aided Civil and Infrastructure Engineering*, 14, 171–185.

PMI. (2009). *A Guide to the Project Management Body of Knowledge*. Fourth edition. PMI. ISBN-10: 1933890517, ISBN-13: 978-1933890517.

Radcliff, D. (2003). Project evaluation and remediation. *Cost Engineering*, 45 (1).

Remenyi, D. (1995). *Effective Measurement and Management of IT Benefits and Cost*. Paper presented at the Management Training and Education Seminar, MTE, Melbourne.

Riel, A. and Lievens, A. (2003). New service development in high tech sectors: A decision-making perspective. In Van Riel, A.C.R. (Ed.) *Effective Decision-making in the High Tech Service Innovation Process*. Doctoral Dissertation. Maastricht: Datawyse/Maastricht University Press.

Rivard, S., Raymond, L. and Verreaul, D. (2006). Resource-based view and competitive strategy: An integrated model of the contribution of information technology to firm performance. *Journal of Strategic Information Systems*, 15, 29–50.

Rosacker, K. and Olson D. (2008). An empirical assessment of IT project selection and evaluation methods in state government. *Project Management Journal*, 39 (1), 49–58.

Saunders, M. et al. (2007). *Research Methods for Business Students*. England: Pearson Education Limited.

Scheuing, E. E. and Johnson, M. E. (1989). New product development and management in financial institutions. *International Journal of Bank Marketing*, 7 (2), 17–21.

Schneider, B. and Bowen, D. E. (1984). New service design, development and implementation and the employee. In George, W. R. and Marshall, C. E. (Eds.) *Developing New Services*. Chicago, IL: American Marketing Association, 82–101.

Scriven, M. (1967). The methodology of evaluation. In Tyler, R. (Ed.) *Perspectives of Curriculum Evaluation*. Chicago, IL: Rand McNally, 39–83.

Scriven, M. (1980). *The Logic of Evaluation*. Edgepress.

Segone (1998). Democratic evaluation, Working paper, UNICEF.

Serafeimidis V. and Smithson S. (2000). Information systems evaluation in practice: a case study of organizational change. *Journal of Information Technology*, 15 (2). 93–105. ISSN 0268-3962.

Shostack, G. L. (1984a). Designing services that deliver. *Harvard Business Review*, January–Febuary, 62, 133–139.

Shostack, G. L. (1984b). Service design in the operating environment. In George, W. R. and Marshall, C. E. (Eds.) *Developing New Services*. Chicago, IL: American Marketing Association, 27–43.

Shtëmbari, E. and Nhung, C. (2013). *Key Criteria in Project Evaluation: A Study of New Service Development*. LAP Lambert Academic Publishing.

Simonsen, J. (2007). Involving top management in IT projects. *Communications of the ACM*, 50, 53–58.

Small, K. A. (1998). Project evaluation. In *Transportation Policy and Economics: A Handbook in Honor of John R. Meyer*. The University of California Transportation Center, Chapter 5.

Stemler, S. (2001). An overview of content analysis. *Practical Assessment, Research & Evaluation*, 7 (17), 137.

Stevens, F., Lawrenz, F. and Sharp, L. (1993). *User Friendly Handbook for Project Management: Science, Mathematics, Engineering and Technology Education*. Washington, DC: National Science Foundation.

Stewart, R. A. (2003). *Life Cycle Management of Information Technology Projects in Construction*. Ph.D. Dissertation: Griffith University.

Stewart, R. A. (2008). A framework for the life cycle management of information technology projects: Project IT. *International Journal of Project Management*, 26, 203–212.

Stewart, R. A. and Mohamed, S. (2002). IT/IS projects selection using multi-criteria utility theory. *Logist Information Management*, 15 (4), 254–270.

Stewart, R. A. and Mohamed, S. (2003). Evaluating the value IT adds to the process of project information management in construction. *Automation in Construction*, 12, 407–417.

Stewart, R. A., Mohamed, S. and Daet, R. (2002). Strategic implementation of IT/IS projects in construction: A case study. *Automat Construct*, 11 (6), 681–694.

Stockdale, R. and Standing, C. (2006). An interpretive approach to evaluating information systems: A content, context, process framework. *European Journal of Operational Research*, 173 (3), 1090–1102.

Storey, C. and Kelly, D. (2001). Measuring the performance of new service development activities. *The Service Industries Journal*, 21 (2), 71–90.

Storey, E. C. (1993). The impact of the new product development project on the success of Financial services. *The Service Industries Journal*, 13 (3), 40–54.

Strauss, A. and Corbin, J. (1990). *Basics of Qualitative Research: Techniques and Procedures for Developing Grounded Theory*. Second edition. New Burry Park, CA: Sage.

Suwardy, T., Ratnatunga, J., Sohal, A. S. and Speight, G. (2003). IT projects: Evaluation, outcomes and impediments. *Benchmarking: An International Journal*, 10 (4), 325–342.

Thomas, D.R.E. (1978). Strategy in different in service businesses. *Harvard Business Review*, July–August, 158–167.

Tukel, O. I. and Walter, O. R. (2001). An empirical investigation of project evaluation criteria. *International Journal of Operations & Production Management*, 21 (3), 400–416.

Turner, J. R. (ed), (1995). The Commercial Project Manager, McGraw-Hill, London, p. 408, ISBN: 0-07-707946-9.

Twiss, B. (1986). *Managing Technological Innovation*. London: Pitman.

Uhl, A. (2000). *The Limits of Evaluation*. Lisbon: European Monitoring Centre for Drugs and Drug Addiction.

Urban, et al. (2014). Evolutionary Evaluation: Implications for Evaluators, Researchers, Practitioners, Funders and the Evidence-Based Program Mandate. *Evaluation and Program Planning*, 45, 127–39.

Vakola, M. (2000). Exploring the relationship between the use of evaluation in business process re-engineering and organisational learning and innovation. *Journal of Management Development*, 19 (10), 812–835.

Valeri, S. G. and Rozenfeld, H. (2004). Improving the flexibility of New Product Development (NPD) through a new quality gate approach. *Society for Design and Process Science*, 8 (3), 17–36.

Van der Aa, W. A. (2002). Realizing innovation in services. *Scandinavian Journal of Management*, 18 (2), 155–171.

Wholey, J., Hatry, H. and Newcomer, K. (1994). *Handbook of Practical Program Evaluation*. San Francisco, CA: Jossey-Bass.

Wiklund, D. and Pucciarelli, J. C. (2009). *Improving IT Project Outcomes by Systematically Managing and Hedging Risk.* IDC report.

Wu and Ong (2008). Management of information technology investment: A framework based on a Real Options and Mean–Variance theory perspective. *Technovation*, 28 (3), 122–134.

Ye, S. and Tiong, R.L.K. (2000). NPV at-risk method in infrastructure project investment evaluation. *Journal of Construction Engineering and Management*, May–June, 227–233.

Yeo, K. and Qiu, F. (2003). The value of management flexibility–A real option approach to investment evaluation. *International Journal of Project Management*, 21 (4), 243–250.

Index